The Amazing World of IDPs in Human Diseases II

The Amazing World of IDPs in Human Diseases II

Editors

Simona Maria Monti
Giuseppina De Simone
Emma Langella

MDPI • Basel • Beijing • Wuhan • Barcelona • Belgrade • Manchester • Tokyo • Cluj • Tianjin

Editors

Simona Maria Monti	Giuseppina De Simone	Emma Langella
Institute of Biostructures and	Institute of Biostructures and	Institute of Biostructures and
Bioimaging (IBB)	Bioimaging (IBB)	Bioimaging (IBB)
National Research Council	National Research Council	National Research Council
(CNR)	(CNR)	(CNR)
Naples	Naples	Naples
Italy	Italy	Italy

Editorial Office
MDPI
St. Alban-Anlage 66
4052 Basel, Switzerland

This is a reprint of articles from the Special Issue published online in the open access journal *Biomolecules* (ISSN 2218-273X) (available at: www.mdpi.com/journal/biomolecules/special_issues/IDPs_II).

For citation purposes, cite each article independently as indicated on the article page online and as indicated below:

LastName, A.A.; LastName, B.B.; LastName, C.C. Article Title. *Journal Name* **Year**, *Volume Number*, Page Range.

ISBN 978-3-0365-3107-6 (Hbk)
ISBN 978-3-0365-3106-9 (PDF)

Contents

About the Editors

Simona Maria Monti The scientific activity of Dr. Monti is focused on chemical biology investigations concerning the structural and functional characterization of proteins and bioactive peptides involved in biomedical pathways and human disease. Her research activity combines several experimental methodologies encompassing molecular biology, X-ray structural investigations, and biomolecular interactions. All this allowed Dr. Monti to establish important and productive collaborations with national and international groups. Dr. Monti is author of more than 122 indexed publications, including *Science, PNAS,* and *JACS,* and has an h-index of 41 (Google Scholar). Her studies gave rise to two international patents. Recently, Dr. Monti co-founded a spinoff company, IMMUNOVEG, dedicated to the green and safe technological production of naturally produced bioactive molecules.

Giuseppina De Simone The research activities of Dr. Giuseppina De Simone are focused on several aspects of structure/function relationships in macromolecules with a biological interest. These studies have been dedicated to a variety of macromolecular systems with different structural complexities and have been carried out by combining X-ray crystallographic methods with several other techniques, including enzyme expression and purification, enzyme functional characterization, molecular modelling, and kinetic and spectroscopic analyses. Dr. De Simone is the author of more than 160 scientific publications of international journals, including *PNAS, Nat Cell Biol, JACS,* and *Chem. Rev,* and has an h-index of 48 (Google Scholar). She has been the principal investigator of several research grants from national and international institutions. Currently, Dr. De Simone is an associate editor of the *Journal of Enzyme Inhibition and Medicinal Chemistry.*

Dr. Emma Langella Dr. Langella graduated in Chemistry summa cum laude at the University of Naples Federico II in 2000 and got her Ph.D. in Chemical Sciences in 2003. Since 2011, she has been a researcher at the Institute of Biostructures and Bioimaging in Naples. The research activity of Dr. Langella is mainly focused on the in silico study of biological systems involved into physiological and pathological processes such as cancer and neurodegenerative diseases. Her main research areas include: (i) the application of molecular dynamics simulation techniques to proteins, protein complexes, macromolecules and oligonucleotides to gain insights into macromolecular conformational preferences through atomic detail; (ii) molecular docking and theoretical binding free energy calculations for computer-aided drug design. Dr. Langella is involved into many national and international research projects and is the co-inventor of a recently deposited international patent. In recent years, Dr. Langella has served as a Guest Editor for several Special Issues in the biomedical field.

Editorial

The Amazing World of IDPs in Human Diseases II

Simona Maria Monti *, Giuseppina De Simone and Emma Langella *

Institute of Biostructures and Bioimaging, CNR, Via Mezzocannone 16, I-80134 Naples, Italy; giuseppina.desimone@cnr.it
* Correspondence: simonamaria.monti@cnr.it (S.M.M.); emma.langella@cnr.it (E.L.)

Citation: Monti, S.M.; De Simone, G.; Langella, E. The Amazing World of IDPs in Human Diseases II. *Biomolecules* **2022**, *12*, 369. https://doi.org/10.3390/ biom12030369

Received: 17 February 2022
Accepted: 21 February 2022
Published: 25 February 2022

Publisher's Note: MDPI stays neutral with regard to jurisdictional claims in published maps and institutional affiliations.

Intrinsically Disordered Proteins (IDPs) lack stable tertiary and secondary structures and are extensively distributed across eukaryotic cells, playing critical roles in cell signaling and regulation [1–3]. IDPs are also frequently associated with the development of diseases such as cancer, cardiovascular diseases, and neurodegenerative diseases [4–7]. For this reason, they have been converted into attractive therapeutic targets, although targeting them is challenging due to their dynamic nature. Indeed, the structural flexibility of IDPs causes difficulties in reliably capturing their heterogeneous structures through conventional experimental methods; thus, new methods and approaches have been developed [8]. This Special Issue includes seven articles from more than 35 scientists around the world working in the amazing field of IDPs. The contributions illustrate the most recent progress in knowledge on IDPs and human diseases.

The Special Issue begins with the article by Atieh and colleagues [9], which describes an interesting investigation on α-synuclein (αSyn) and DJ-1, an antioxidative protein that plays a critical role in Parkinson's disease (PD) pathology. Through nuclear magnetic resonance (NMR) spectroscopy integrated with atomic force microscopy (AFM) in solution, the authors characterized the interaction of DJ-1 with glycated N-terminally acetylated-αSyn (glyc-ac-αSyn). The obtained results show that DJ-1 interacts with glycated and native ac-αSyn through the catalytic triad and establish that the oxidation state of the catalytic cysteine is imperative for binding. A mechanism of action by which DJ-1 interacts with N-terminally acetylated-αSyn oligomers, preventing their interaction with glyc-ac-αSyn monomers, was proposed. The relevance of these results within PD pathology shows how DJ-1 function in chaperoning αSyn may prevent the rapid accumulation of aggregated αSyn within the cell, which may enable proper clearance mechanisms from the cell and reduce the effects of neurodegeneration. Therapeutics targeting the effects of glycation in conjunction with maintaining proper DJ-1 function may mitigate neurodegeneration and diminish the symptoms of PD.

Another remarkable study within this Special Issue was carried out by Rizzuti and coworkers [10] on nuclear protein 1 (NUPR1), which is a small, highly basic ID protein of 82 residues that localizes throughout the whole cell, and is involved in the development and progression of several tumors. Based on previous results, the authors designed and synthetized nine derivatives, starting from lead compound ZZW-115, which were then investigated through biophysical and cellular experiments. Interestingly, the authors highlight how a more favorable binding affinity does not necessarily correlate with biological effects, underlining the importance of having a subtle compromise between increasing drug affinity and altering protein function, in addition to other properties such as solubility, crowding, membrane permeation, cellular efflux and cellular metabolism.

Cardone and co-workers thoroughly investigated phosphoprotein P of Mononegavirales (MNV) [11], which is an essential co-factor of the viral RNA polymerase L and whose prime function is to recruit L to the ribonucleocapsid composed of the viral genome encapsidated by the nucleoprotein N. The authors investigated the dynamic behavior of $P_{C\alpha}$, a domain that is C-terminal to the small oligomerization domain (P_{OD}) and constitutes the respiratory syncytial virus L-binding region together with P_{OD}. By using small phosphoprotein fragments centered on or adjacent to P_{OD}, a structural picture of the P_{OD}–$P_{C\alpha}$ region in solution was gained, evidencing how small molecules are able to modify the dynamics of $P_{C\alpha}$. This observed structural plasticity of the $P_{C\alpha}$ domain may play a crucial role for the functional viral polymerase, which needs more investigations.

The paper by Ortega-Alarcon and colleagues [12] investigated methyl-CpG binding protein 2 (MeCP2), a multidomain IDP associated with neuronal development and maturation. In particular, the authors focused their attention on MBD, one of the key domains in MeCP2 responsible for DNA recognition, and its two flanking disordered domains, NTD and ID. It was demonstrated that both the disordered domains—NTD and ID—unequivocally stabilize the MBD domain against thermal and chemical denaturation and that NTD-MBD-ID differs functionally and structurally from MBD. The authors also highlight how disorder in proteins may be considered a pervasive feature that is even more important in multidomain IDPs with a complex conformational and multifunctional landscape.

The structural features of FOXO3 were analyzed by Weinzierl [13] through multiple independent molecular dynamics simulations of models of full-length FOXO3 bound to DNA, using both implicit and explicit solvation conditions. FOXO3, belonging to the 'forkhead box O' gene family, is of considerable interest in many therapeutically relevant areas, such as tumor therapy and longevity research. The obtained results provide atomistic models for an extended structure of FOXO3 when bound to DNA, showing that the two 'linker' regions immediately adjacent to the DNA-binding domain are present in a highly extended conformation, likely due to electrostatic repulsion of the domains connected by the linkers. The study sheds light on previously unrecognized structural properties of FOXO3 and introduces a new graphical method of general use that is particularly helpful for studying and visualizing the structural diversity of IDPs.

Bokor and Tantos [14] studied two different IDP interaction systems to gain information about the bonds holding the protein associations together using wide-line 1H NMR. One system consisted of wild type and mutant α-synuclein (αS) in the forms of oligomers and amyloids and the other system was the complex between the intrinsically disordered (IDP) thymosin-β4 (Tβ4) and the cytoplasmic domain of stabilin-2 (stabilin CTD), which is involved in the phagocytosis of apoptotic cells. The study provides insights into the intermolecular bonds that contribute to the formation of αS oligomer and amyloid aggregates. Moreover, the authors revealed information on the molecular background of the fuzzy complexes between thymosin-β4 and stabilin-2 CTD.

The work presented by Raut and colleagues [15] investigated Par-4 (Prostate apoptosis response-4), a predominantly intrinsically disordered protein, acting as a tumoural suppressor that is capable of selectively inducing apoptosis in cancer cells while leaving healthy cells unaffected. The authors performed experimental studies, employing circular dichroism spectroscopy and dynamic light scattering to assess the effects of various monovalent and divalent salts upon the conformation of cl-Par-4 in vitro. The obtained results clarify the different roles of cations and anions in influencing Par-4 structure and indicate that the SAC domain of the protein, which is the region of Par-4 indispensable for its apoptotic function, is likely to be helical in cl-Par-4 under the studied high salt conditions.

In summary, the papers collected in this Special Issue unveil novel aspects related to the wide world of IDPs. Using a variety of methods, including biochemical, spectroscopic and computational techniques, they represent a step forward in the study and characterization of many IDPs involved in human diseases, with a focus on conformational features, environmental effects, recognition mechanisms and targeting.

Conflicts of Interest: The authors declare no conflict of interest.

References

1. Darling, A.L.; Uversky, V.N. Intrinsic Disorder and Posttranslational Modifications: The Darker Side of the Biological Dark Matter. *Front. Genet.* **2018**, *9*, 158. [CrossRef] [PubMed]
2. Buonanno, M.; Coppola, M.; Di Lelio, I.; Molisso, D.; Leone, M.; Pennacchio, F.; Langella, E.; Rao, R.; Monti, S.M. Prosystemin, a prohormone that modulates plant defense barriers, is an intrinsically disordered protein. *Protein Sci.* **2018**, *27*, 620–632. [CrossRef] [PubMed]
3. Sun, X.; Jones, W.T.; Rikkerink, E.H.A. GRAS proteins: The versatile roles of intrinsically disordered proteins in plant signalling. *Biochem. J.* **2012**, *442*, 1–12. [CrossRef] [PubMed]
4. Ayyadevara, S.; Ganne, A.; Balasubramaniam, M.; Shmookler Reis, R.J. Intrinsically disordered proteins identified in the aggregate proteome serve as biomarkers of neurodegeneration. *Metab. Brain Dis.* **2022**, *37*, 147–152. [CrossRef] [PubMed]
5. Martinelli, A.H.S.; Lopes, F.C.; John, E.B.O.; Carlini, C.R.; Ligabue-Braun, R. Modulation of Disordered Proteins with a Focus on Neurodegenerative Diseases and Other Pathologies. *Int. J. Mol. Sci.* **2019**, *20*, 1322. [CrossRef] [PubMed]
6. Langella, E.; Buonanno, M.; De Simone, G.; Monti, S.M. Intrinsically disordered features of carbonic anhydrase IX proteoglycan-like domain. *Cell. Mol. Life Sci.* **2021**, *78*, 2059–2067. [CrossRef] [PubMed]
7. Langella, E.; Buonanno, M.; Vullo, D.; Dathan, N.; Leone, M.; Supuran, C.T.; De Simone, G.; Monti, S.M. Biochemical, biophysical and molecular dynamics studies on the proteoglycan-like domain of carbonic anhydrase IX. *Cell. Mol. Life Sci.* **2018**, *75*, 3283–3296. [CrossRef] [PubMed]
8. Santofimia-Castaño, P.; Rizzuti, B.; Xia, Y.; Abian, O.; Peng, L.; Velázquez-Campoy, A.; Neira, J.L.; Iovanna, J. Targeting intrinsically disordered proteins involved in cancer. *Cell. Mol. Life Sci.* **2020**, *77*, 1695–1707. [CrossRef] [PubMed]
9. Atieh, T.B.; Roth, J.; Yang, X.; Hoop, C.L.; Baum, J. DJ-1 Acts as a Scavenger of α-Synuclein Oligomers and Restores Monomeric Glycated α-Synuclein. *Biomolecules* **2021**, *11*, 1466. [CrossRef]
10. Rizzuti, B.; Lan, W.; Santofimia-Castaño, P.; Zhou, Z.; Velázquez-Campoy, A.; Abián, O.; Peng, L.; Neira, J.L.; Xia, Y.; Iovanna, J.L. Design of Inhibitors of the Intrinsically Disordered Protein NUPR1: Balance between Drug Affinity and Target Function. *Biomolecules* **2021**, *11*, 1453. [CrossRef]
11. Cardone, C.; Caseau, C.M.; Bardiaux, B.; Thureaux, A.; Galloux, M.; Bajorek, M.; Eléouët, J.F.; Litaudon, M.; Bontems, F.; Sizun, C. A Structural and Dynamic Analysis of the Partially Disordered Polymerase-Binding Domain in RSV Phosphoprotein. *Biomolecules* **2021**, *11*, 1225. [CrossRef] [PubMed]
12. Ortega-Alarcon, D.; Claveria-Gimeno, R.; Vega, S.; Jorge-Torres, O.C.; Esteller, M.; Abian, O.; Velazquez-Campoy, A. Stabilization effect of intrinsically disordered regions on multidomain proteins: The case of the methyl-cpg protein 2, mecp2. *Biomolecules* **2021**, *11*, 1216. [CrossRef] [PubMed]
13. Weinzierl, R.O.J. Molecular dynamics simulations of human foxo3 reveal intrinsically disordered regions spread spatially by intramolecular electrostatic repulsion. *Biomolecules* **2021**, *11*, 856. [CrossRef] [PubMed]
14. Bokor, M.; Tantos, Á. Protein–Protein connections—Pligomer, amyloid and protein complex—By wide line1 h nmr. *Biomolecules* **2021**, *11*, 757. [CrossRef]
15. Raut, K.K.; Ponniah, K.; Pascal, S.M. Structural Analysis of the cl-Par-4 Tumor Suppressor as a Function of Ionic Environment. *Biomolecules* **2021**, *11*, 386. [CrossRef] [PubMed]

biomolecules

MDPI

Article

DJ-1 Acts as a Scavenger of α-Synuclein Oligomers and Restores Monomeric Glycated α-Synuclein

Tamr B. Atieh, Jonathan Roth, Xue Yang, Cody L. Hoop and Jean Baum

Department of Chemistry and Chemical Biology, Rutgers University, Piscataway, NJ 08854, USA; tamr@chem.rutgers.edu (T.B.A.); jr1214@chem.rutgers.edu (J.R.); yangxuechemistry@hotmail.com (X.Y.); cody.hoop@rutgers.edu (C.L.H.)
* Correspondence: jean.baum@rutgers.edu

Citation: Tamr B. Atieh, Jonathan Roth, Xue Yang, Cody L. Hoop and Jean Baum DJ-1 Acts as a Scavenger of α-Synuclein Oligomers and Restores Monomeric Glycated α-Synuclein. *Biomolecules* **2021**, *11*, 1466. https://doi.org/10.3390/biom11101466

Academic Editors: Simona Maria Monti, Giuseppina De Simone and Emma Langella

Received: 26 August 2021
Accepted: 1 October 2021
Published: 6 October 2021

Publisher's Note: MDPI stays neutral with regard to jurisdictional claims in published maps and institutional affiliations.

Abstract: Glycation of α-synuclein (αSyn), as occurs with aging, has been linked to the progression of Parkinson's disease (PD) through the promotion of advanced glycation end-products and the formation of toxic oligomers that cannot be properly cleared from neurons. DJ-1, an antioxidative protein that plays a critical role in PD pathology, has been proposed to repair glycation in proteins, yet a mechanism has not been elucidated. In this study, we integrate solution nuclear magnetic resonance (NMR) spectroscopy and liquid atomic force microscopy (AFM) techniques to characterize glycated N-terminally acetylated-αSyn (glyc-ac-αSyn) and its interaction with DJ-1. Glycation of ac-αSyn by methylglyoxal increases oligomer formation, as visualized by AFM in solution, resulting in decreased dynamics of the monomer amide backbone around the Lys residues, as measured using NMR. Upon addition of DJ-1, this NMR signature of glyc-ac-αSyn monomers reverts to a native ac-αSyn-like character. This phenomenon is reversible upon removal of DJ-1 from the solution. Using relaxation-based NMR, we have identified the binding site on DJ-1 for glycated and native ac-αSyn as the catalytic pocket and established that the oxidation state of the catalytic cysteine is imperative for binding. Based on our results, we propose a novel mechanism by which DJ-1 scavenges glyc-ac-αSyn oligomers without chemical deglycation, suppresses glyc-ac-αSyn monomer–oligomer interactions, and releases free glyc-ac-αSyn monomers in solution. The interference of DJ-1 with ac-αSyn oligomers may promote free ac-αSyn monomer in solution and suppress the propagation of toxic oligomer and fibril species. These results expand the understanding of the role of DJ-1 in PD pathology by acting as a scavenger for aggregated αSyn.

Keywords: α-synuclein; glycation; DJ-1; Parkinson's disease; protein–protein interactions; nuclear magnetic resonance spectroscopy; atomic force microscopy

1. Introduction

As the population shifts towards an aging society, it is imperative to understand the effect of aging on neurodegenerative diseases. One result of aging is the body's inability to mitigate the harmful impacts of glucose metabolism, which produces reactive oxygen species (ROS) and reactive aldehyde species [1,2]. A direct consequence of this aldehyde formation is the non-enzymatic chemical ligation of sugar aldehydes to the side chains of susceptible proteins in the formation of advanced glycation end-products (AGEs) [3]. Protein glycation has been linked to multiple degenerative diseases such as Parkinson's disease (PD) [4], Alzheimer's disease [5], Huntington's disease [6], diabetes [7], and atherosclerosis [8]. Glycation of amyloidogenic proteins associated with these diseases has been shown to induce their formation of toxic oligomers that are unable to be cleared by the cell [9,10]. A build-up of glycated protein in neurons adds another challenge against combating debilitating neurodegenerative diseases. Suppressing the aggregation of these toxic glycated species may be a viable approach toward alleviating the effects of aging-related neurodegeneration.

Aggregation of the intrinsically disordered protein α-synuclein (αSyn) is associated with multiple neurodegenerative diseases including PD, multiple system atrophy, and Lewy Body Dementia [11,12] and leads to the formation of Lewy bodies in the substantia nigra of neurons. The misfolding and aggregation of αSyn is very complex and involves a self-association of monomers, the development of heterogenous oligomeric species that vary in size and toxicity, and progression into fibrils that are incorporated into Lewy bodies [13,14]. Although the mechanism and culprit species for PD pathogenicity has not been elucidated, ample research has highlighted the toxicity of αSyn oligomers. αSyn oligomers have been observed at elevated levels in the brains of PD patients and transgenic mouse and Drosophila models associated with disease [15–17]. In vivo and in vitro studies have shown that αSyn oligomers formed in multiple conditions are toxic to neurons [18–21]. Many of these oligomers are able to propagate amyloid formation [22]. These smaller oligomeric species and fragmented fibrils are able to spread cell-to-cell and seed amyloid formation in a prion-like manner in ways long, mature amyloid fibrils cannot [18]. Therefore, αSyn oligomers may serve as an early interventional target against amyloid propagation.

Glycation of αSyn, a result of aging, leads to the formation of insoluble protein plaques and toxic oligomers that do not form fibrils and are unable to be cleared by the cell and may cause neuronal death [10,23–26]. AGEs, including glycated-αSyn (glyc-αSyn), are found at significantly higher levels in neurons of PD patients than in healthy neurons [4,10,27]. A potent glycating agent in cells is methylglyoxal (MGO), which is produced as a byproduct of glucose metabolism [28]. MGO glycates lysine, arginine, and cysteine side chains of proteins and has been shown to glycate specific lysines in the N-terminus of αSyn [10,29,30]. MGO-mediated glyc-αSyn has been postulated to play a role in Lewy Body disease and can co-aggregate with native αSyn to suppress amyloid formation [31,32]. Compounding this issue, glyc-αSyn is unable to undergo proper degradation via normal cellular pathways such as ubiquitination, cellular autophagy, or synaptic transmission [10]. Gaining a molecular understanding of ways to interfere with an accumulation of these toxic oligomeric species of glyc-αSyn is a crucial step for relieving aging effects in PD patients. Therapeutics targeting glycation pathways or that chaperone harmful aggregates may enhance the neuroprotective pathways within the cell.

Central to mitigation of the adverse effects of aging glucose metabolism lies the antioxidative protein DJ-1 [33]. DJ-1 exists as a homodimeric protein in solution and is crucial for neuronal protection against oxidative stress [34–36]. The oxidation state of the highly conserved cysteine at position 106, a residue within the catalytic triad, is imperative to the functionality of DJ-1 [37,38]. DJ-1 can be translocated to the mitochondria to scavenge for ROS produced from glucose metabolism within the cell [39], lessening oxidative stress [40]. DJ-1 has been suggested to act as a deglycase, which may mitigate the effects of glucose metabolism [41]. Recent in vitro studies show that DJ-1 can deglycate chemically glycated lysine, arginine, and cysteine amino acids as well as repair glutathione following an MGO attack [42]. However, DJ-1's repair mechanism for larger proteins remains inconclusive [43,44].

DJ-1 has been linked to early onset PD [45,46]. The DJ-1 familial mutant L166P causes reduced stability of the homodimer and leads to early onset PD [47]. DJ-1 and αSyn are thought to colocalize within neurons and may directly interact with one another to modulate αSyn aggregation kinetics [48–50]. Indeed, DJ-1, with Cys106 in the sulfinic acid form, has already been shown to act as a chaperone for native αSyn and inhibit protofibril and amyloid formation in vivo and in vitro [51,52], and we show further data to support these findings (Figure S1). More recently, Kumar et al. concluded that DJ-1 directly interacts with and remodels mature αSyn fibrils and produces species that are more toxic to SH-SY5Y cells than fibrils themselves [53]. In addition, DJ-1 deficiency can lead to an accumulation of αSyn in neurons, while DJ-1 overexpression leads to a decrease in αSyn levels [54]. Several attempts to ascertain the direct interactions between αSyn and DJ-1 have been unsuccessful [55,56]. DJ-1 does not exhibit tight binding to αSyn but may have transient binding as seen through bimolecular fluorescence complementation and coimmunoprecipitation [49]. Given their co-localization, it has been speculated that DJ-1 may deglycate glyc-αSyn to repair toxic formation of AGEs in neurons [57]. However, no direct evidence has been presented. A molecular view of the impact of DJ-1 on αSyn glycation in aging PD patients would aid in the design of therapeutics against detrimental effects of αSyn glycation.

Here, through the integration of solution nuclear magnetic resonance (NMR) spectroscopy and atomic force microscopy (AFM) in solution, we characterize glyc-αSyn and its interactions with DJ-1. Throughout our study, we use N-terminally acetylated αSyn (ac-αSyn), the ubiquitous post-translational modification found in LBs, which represents the physiologically relevant form of αSyn [58,59]. N-terminal acetylation affects the conformational ensemble of the monomer and the aggregation kinetics [60,61]. Therefore, this modification is significant in its molecular interactions with other αSyn molecules or other proteins. Relaxation-based NMR illuminates the unique transient glyc-ac-αSyn monomer–oligomer binding events and suggests that the presence of DJ-1 reduces these interactions, which are recovered upon removal of DJ-1. We determine that DJ-1 interacts with glycated and native ac-αSyn through the catalytic triad and establish that the oxidation state of the catalytic cysteine is imperative for binding. Supported by AFM imaging in solution, we propose a mechanism by which DJ-1 interacts with glyc-ac-αSyn oligomers, preventing their interaction with glyc-ac-αSyn monomers, leaving a higher population of free monomers in solution. Within PD pathology, DJ-1's function in chaperoning αSyn may prevent the rapid accumulation of aggregated αSyn within the cell, which may enable proper clearance mechanisms from the cell and reduce the effects of neurodegeneration. Therapeutics targeting the effects of glycation in conjunction with maintaining proper DJ-1 function may successfully mitigate neurodegeneration and diminish the symptoms of PD.

2. Materials and Methods

2.1. Protein Expression and Purification of Acetylated α-Synuclein

All α-synuclein, including glycated α-synuclein, used in this work is N-terminally acetylated. Acetylated α-synuclein was expressed and purified as previously described [61]. To acetylate, αSyn and NATB plasmids were co-transformed into BL21(DE3) *E. coli* cells. Cell cultures were grown in either nutrient-rich Luria Broth (LB) or minimal M9 media supplemented with ^{15}N-ammonium chloride and/or ^{13}C-glucose for isotopic enrichment for NMR. Cell cultures (1 L) were grown at 37 °C with shaking until they reached an OD_{600} of 0.6–0.8, at which point 1 mM IPTG was added to induce expression and incubated at 37 °C with shaking for 4 h. The cell cultures were then spun down at 4.5k rpm for 30 min and the pellet was resuspended in 20 mL of phosphate buffered saline (PBS), pH 7.4, and then homogenized three times at 10–15 k psi. The cell lysates were spun down at 20 k rpm for 30 min, and 10 mg/mL of streptomycin sulfate was added to the supernatant and mixed for 15 min at 4 °C. Once completed, the suspension was spun down again at 20 k rpm for 30 min. The supernatant was collected, mixed with 0.361 g/mL of ammonium sulfate, and incubated at 4 °C for 1 h to precipitate the proteins. The mixture was again spun down for 30 min at 20 k rpm and the supernatant was discarded. The protein pellet was dissolved in 15 mL of PBS buffer and double boiled for 20 min and allowed to cool to room temperature. The supernatant was collected after centrifugation and dialyzed against 15 mM Tris buffer overnight at 4 °C. The protein solution was filtered through a 0.22-micrometer filter and passed through an anion exchange column (Hitrap Q HP, GE Lifesciences, Piscataway, NJ, USA). Ac-αSyn was eluted with a 250 mM NaCl gradient. All protein-containing fractions (as assessed by UV_{280}) were collected and dialyzed with four buffer changes against 15 mM ammonium bicarbonate and lyophilized. Protein purity was assessed via SDS-PAGE and electrospray ionization mass spectrometry (ESI-MS) to ensure proper acetylation. Lyophilized ac-αSyn powder was stored at −20 °C.

2.2. Expression and Purification of DJ-1

A plasmid encoding human DJ-1 with a C-terminal His-tag was purchased from Addgene (#51488). The plasmid was transformed into BL21(DE3) *E. coli* cells. Cell cultures were grown in LB or M9 media (with isotopic enrichment for NMR) at 37 °C with shaking and allowed to reach an OD_{600} of 0.6–0.8. Expression was induced with 1 mM IPTG at 20 °C overnight. Cells were harvested and resuspended in 50 mM sodium phosphate buffer (pH 8), 300 mM NaCl, 20 mM imidazole, 1 mM DTT, and 5% glycerol. These cells were homogenized three times at 10–15 k psi and centrifuged at 20 k rpm for 30 min to remove any cellular debris. The cell lysate was filtered through a 0.22-micron filter and passed over a His trap column equilibrated with 50 mM sodium phosphate buffer (pH 8), 300 mM NaCl, 20 mM imidazole, and 5% glycerol. DJ-1 was eluted from the column with 200 mM imidazole. DJ-1 fractions were collected and dialyzed against PBS overnight at 4 °C. For preparations for size exclusion chromatography (SEC), DJ-1 was concentrated using a 3 kD filter and filtered through a 0.22-micron filter. The sample was then passed over a Superdex 200 Increase 10/300 GL SEC column that was equilibrated with PBS. DJ-1 fractions were collected and promptly oxidized. DJ-1 purity was assessed via ESI-MS and SDS PAGE-gel. The C-terminal his-tag was retained.

2.3. Oxidation of DJ-1

DJ-1 at 400 µM dimer concentration was incubated with 800 µM H_2O_2 at 4 °C overnight, and then buffer exchanged to PBS at pH 7.4 to remove any excess H_2O_2. The concentration of dimeric DJ-1 was measured using A_{280} with a molar extinction coefficient of 8400 $M^{-1}cm^{-1}$ (4200 $M^{-1}cm^{-1}$ monomer). Unless otherwise stated, DJ-1 is oxidized to the sulfinic acid form before experiments, as confirmed using ESI-MS.

2.4. Glycation of Acetylated α-Synuclein

Lyophilized ac-αSyn was dissolved in PBS at pH 7.4 and subsequently passed through a 100 kD filter to remove aggregates and diluted to a final concentration of 50 µM monomeric ac-αSyn in PBS. The protein sample was incubated with 50 mM MGO at 37 °C for 24 h. All MGO was removed via dialysis with four buffer exchanges with PBS. To prepare samples for experiments, glycated ac-αSyn (glyc-ac-αSyn) was passed through a 100 kD filter, washed with PBS, and concentrated with a 3 kD filter. Protein concentrations of monomeric glyc-ac-αSyn were assessed via a BCA assay.

2.5. Reaction Conditions for DJ-1 and Glycated ac-α-Synuclein

For experiments on glyc-ac-αSyn in the presence of DJ-1 (+DJ-1), 100 µM glyc-ac-αSyn was incubated with 20 µM DJ-1 for 1 h at 37 °C in PBS at pH 7.4. For experiments on glyc-ac-αSyn after the removal of DJ-1 (–DJ-1), DJ-1 was removed by passing the glyc-ac-αSyn+DJ-1 solution over a His-trap to remove the His-tagged DJ-1. The glyc-ac-αSyn incubated with DJ-1 was assessed for purity by SDS-PAGE gel to ensure proper removal of DJ-1, and final concentration of glyc-ac-αSyn was determined using a BCA assay.

2.6. NMR ^1H–^{15}N 2D Correlation Spectra and ^{15}N-R_2 Experiments

NMR experiments were performed on 250 µM uniformly ^{15}N-labelled native or glycated ac-αSyn at 15 °C. Lyophilized native or glycated ac-αSyn powder was dissolved in PBS buffer, pH 7.4 and filtered through a 100 kD centrifugal filter to remove large aggregates. Protein was concentrated with a 3 kD centrifugal filter and concentrations were measured via a BCA assay. NMR experiments on uniformly ^{15}N-labelled DJ-1 (500 µM monomer equivalent) in PBS buffer, pH 7.4 were performed at 25 °C. All experiments were performed at 700 MHz ^1H Larmor frequency.

For ac-αSyn samples, ^{15}N-transverse relaxation rates (R_2) were measured from a series of heteronuclear single quantum coherence (HSQC)-based 2D ^1H–^{15}N correlation spectra implementing the Carr-Purcell-Meiboom-Gill (CPMG) pulse sequence with varying relaxation delays: 8, 16, 32, 64, 72, 128, 160, 192, 256, 288, 320, 352, and 384 ms. ^{15}N-

R_2 rates of ^{15}N-DJ-1 were measured from a series of transverse-relaxation optimized spectroscopy (TROSY) ^{1}H–^{15}N correlation spectra using the CPMG pulse sequence with varying relaxation delays: 0, 8, 16, 24, 32, 40, 48, 56, 64, 72, 80 ms.

All titrations involve acquiring a ^{1}H–^{15}N 2D NMR spectrum of the ^{15}N-labelled protein followed by addition of the natural abundance protein. After each subsequent protein addition, the sample was incubated at 37 °C for one hour to help facilitate the DJ-1–ac-αSyn reaction before acquisition.

^{1}H–^{15}N-HSQC, ^{1}H–^{15}N-TROSY, and ^{15}N-R_2 experiments were processed via NMR-Pipe [62] and analyzed in Sparky [63] software. ^{15}N-R_2 rates were measured by fitting a single exponential decay function to the peak intensities of the decay curves using the relaxation peak heights (rh) program in Sparky.

2.7. Thioflavin T Assay

Lyophilized native or glycated ac-αSyn was dissolved in PBS, passed through a 100 kD filter to remove large aggregates, and concentrated and washed using a 3 kD centrifugal filter. Thioflavin T (ThT) reactions consisted of 70 μM native or glycated ac-αSyn with 20 μM ThT in PBS. Where indicated, DJ-1 was added to samples at 140 μM. A total of 100 μL of the samples were aliquoted into clear-bottom 96-well plate with one Teflon bead to each reaction, sealed with Axygen sealing tape (Corning), and shaken at 600 rpm at 37 °C in a POLARstar Omega fluorimeter (BMG Labtech) for over 100 h. Fluorescence was monitored every 33 min.

2.8. Thioflavin T Seeding Experiments

Fibril seeds were prepared as previously described [64]. In brief, 10 mg/mL lyophilized native ac-αSyn was dissolved in PBS in a microcentrifuge tube and shaken at 300 rpm for 5 days without a Teflon bead. The resulting solution was then centrifuged at 10 k rpm for 30 min and resuspended with PBS twice to ensure all monomeric ac-αSyn was removed. Fibril concentration was assessed by dissolving an aliquot of fibrils in 8 M guanidinium hydrochloride and measuring A_{280}. In seeded ThT assays, 1 μM of fibril seeds were added to 70 μM of monomeric native or glycated ac-αSyn, 20 μM ThT, in PBS at pH 7.4. Fibril growth was monitored as a function of ThT fluorescence measured every 33 min at 37 °C under quiescent conditions (without shaking).

2.9. UV–vis Spectroscopy

UV–vis wavelength scans of 100 μM native or glycated ac-αSyn in the presence or absence of DJ-1 in PBS, pH 7.4 were acquired in a 1-centimeter quartz cuvette using PBS as the blanking buffer. Initial concentrations were determined using a BCA assay. Absorbance was measured at variable wavelengths from 250–500 nm in increments of 0.5 nm on a UV–vis spectrophotometer.

2.10. Liquid AFM Imaging

AFM images were acquired on a Cypher ES AFM (Asylum Research) using PNP-DB tips with a nominal spring constant of ≈0.5 N/m and drive frequency of ≈67 kHz. For sample preparation, mica was first treated in an aminopropyl silatrane (APS) solution (67 μM) for 30 min to functionalize the surface and then washed thoroughly with ultrapure water and PBS. The samples were then deposited onto the surface (50 μL droplet, all samples had a concentration of 10 μg/mL) and allowed to bind for 20 min at room temperature. The surface was then washed again with PBS and placed into the AFM for imaging. Care was taken to ensure the samples were never allowed to dry. All images were taken at room temperature in standard tapping mode with a resolution of 256 × 256 pixels. Additionally, blueDrive (Asylum Research, Oxford Instruments) photothermal excitation was utilized to ensure high quality imaging in liquid conditions.

Images obtained were processed using the "Particle Analysis" function in the Asylum Research AFM software, which yielded heights for all particles in the images. These values were then grouped by sample and exported into MATLAB (R2020b). A one-way analysis of variance (ANOVA) was then performed on the grouped data to determine significant differences.

3. Results

3.1. Lysine-Rich N-Terminus and NAC of ac-αSyn Are Most Susceptible to Glycation Effects

Glycated ac-αSyn (glyc-ac-αSyn) was produced from the reaction of ac-αSyn with MGO. The resulting glyc-ac-αSyn is distinct from native ac-αSyn in chemical composition, as assessed using the UV–Vis absorbance profile, and aggregation characteristics, evaluated from size exclusion chromatography (SEC) and thioflavin T (ThT) fluorescence) (Figure S2). The SEC chromatogram shows the presence of small oligomers (~13–16 mL elution) that form due to the glycation of ac-αSyn and are not present in native ac-αSyn. In order to determine the residue-specific effects of glycation on ac-αSyn, we used solution NMR to investigate residue specific perturbations to structure and dynamics of ^{15}N-ac-αSyn monomers in solution upon glycation with MGO. Native αSyn consists of 15 lysines that are susceptible to MGO-mediated glycation, several of which are part of the imperfect KTKEGV repeats that are concentrated in the N-terminus (Figure 1A). The ^{1}H–^{15}N HSQC (heteronuclear single quantum coherence) spectrum of ^{15}N-glyc-ac-αSyn (Figure 1B) shows that observed resonances have a significant peak overlap with the native form of ac-αSyn. The ^{15}N-glyc-ac-αSyn sample contains a mixture of monomers and non-isolatable oligomers observed in SEC. However, the oligomers likely tumble too slowly in solution to be detected by solution NMR. The considerable resonance overlap with native ac-αSyn indicates that we are observing monomeric glyc-ac-αSyn in the ^{1}H–^{15}N HSQC and that it maintains an intrinsically disordered structure and a similar conformational ensemble to native ac-αSyn. However, substantial peak intensity losses with increased ^{15}N-transverse relaxation rates (R_2) are observed in the lysine-rich N-terminal and the non-amyloidβ component (NAC) regions of glyc-ac-αSyn (Figure 1C) and small chemical shift perturbations are observed in C-terminal residues (Figure 1D,E). The significant broadening observed in N-terminal and NAC residues suggests that this region is undergoing intermediate exchange, which may arise due to the chemical glycation, conformational changes, and/or interactions between the glyc-ac-αSyn monomers and undetectable oligomers in solution that are induced by the glycation reaction. A slight increase in ^{15}N-R_2 values is observed in the C-terminus of ^{15}N-glyc-ac-αSyn only up to residue 106, four residues past the most C-terminal lysine, K102, with less substantial peak intensity losses near the C-terminal lysines compared to the N-terminal and NAC regions. However, the observation of chemical shift perturbations in the C-terminus indicates that this region experiences fast exchange, which may occur due to conformational changes in the protein or weak interactions. Thus, the N-terminal and NAC regions are most susceptible to glycation effects caused by the MGO reaction.

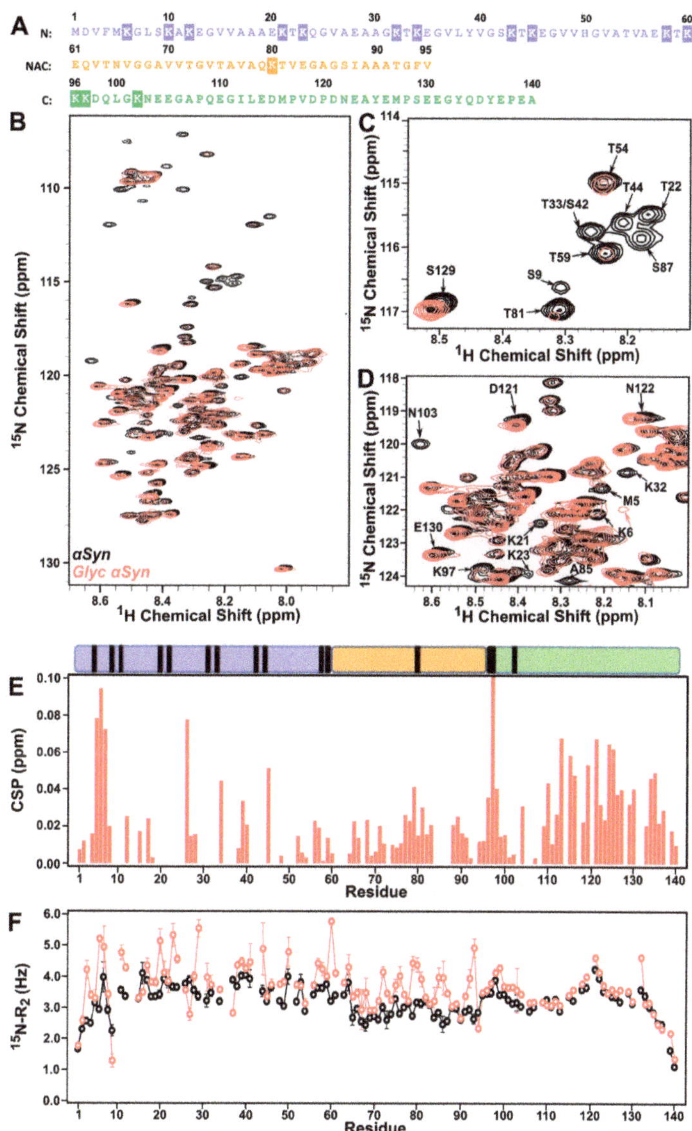

Figure 1. Residue-specific dynamic differences between native ac-αSyn and glyc-ac-αSyn. (**A**) Primary sequence of ac-αSyn, highlighting the lysines as potential glycation sites for MGO (purple—N-terminal, orange—NAC, and green—C-terminal domains). (**B**) ^{1}H–^{15}N HSQC spectra of native ac-αSyn (black) and glyc-ac-αSyn (red) indicating the intrinsically disordered structure of ac-αSyn even upon glycation. (**C**–**D**) Zoomed-in regions of the ^{1}H–^{15}N HSQC shown in (**B**) of (**C**) the serine and threonine region, which shows a dramatic drop in peak intensities upon glycation, and (**D**) a region that highlights chemical shift perturbations (CSPs) of C-terminal residues upon glycation. (**E**) The chemical shift perturbations of glyc-ac-αSyn relative to native ac-αSyn show that perturbations are localized in the early N-terminus and C-terminus. The three domains of ac-αSyn are shown at the top and color-coded as in (**A**). Lysines are highlighted with black boxes. (**F**) The ^{15}N transverse relaxation rates (^{15}N-R$_2$) of native ac-αSyn (black) and glyc-ac-αSyn (red) show decreased dynamics in the N-terminus and NAC, near lysine residues. C-terminal residues show indistinguishable ^{15}N-R$_2$ values between native and glyc-ac-αSyn. Error bars are determined from the fitting errors of the single exponential decay fits. All spectra were acquired on 250 μM ac-αSyn in PBS at 15 °C.

3.2. DJ-1 Restores Native-like Character to Glyc-ac-αSyn

DJ-1 has been shown to mitigate the effects of MGO on free amino acids as a glyoxalase [33]; however, mechanistic details of how DJ-1 impacts glycation of larger proteins remains controversial [44,65]. Although, there is some evidence that it may deglycate small compounds and free amino acids [42,66]. In order to determine the residue-specific effect of DJ-1 on glyc-ac-αSyn, we monitored ^{15}N-glyc-ac-αSyn using NMR upon addition of DJ-1. Upon incubation of ^{15}N-glyc-ac-αSyn with a 1:1 molar ratio of DJ-1, surprisingly, residue specific chemical shifts and peak intensities of ^{15}N-glyc-ac-αSyn in ^{1}H–^{15}N HSQC spectra are now similar to those of native ac-αSyn (Figure 2A, red and Figure S3), and ^{15}N-R$_2$ values are also indistinguishable from ^{15}N-native ac-αSyn (Figure 2B). This suggests that the interaction of DJ-1 with glyc-ac-αSyn restores structural and dynamic characteristics of native ac-αSyn to glyc-ac-αSyn either through chemical deglycation, or by suppressing conformational exchange of the glyc-ac-αSyn protein.

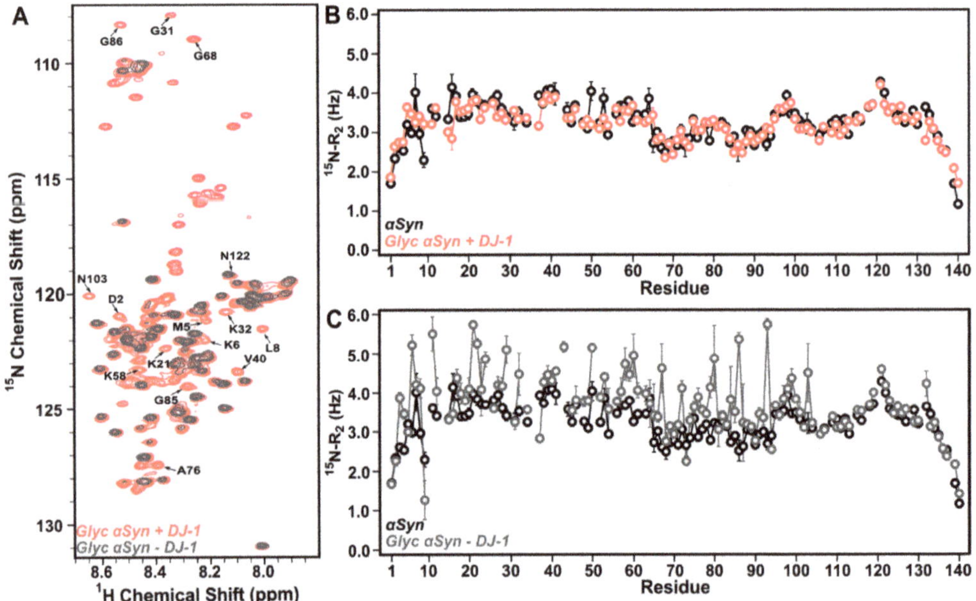

Figure 2. DJ-1 restores native-like character to glyc-ac-αSyn. (**A**) ^{1}H–^{15}N HSQC spectra of 250 μM ^{15}N-glyc-ac-αSyn in the presence of 250 μM DJ-1 (red) or upon removal of DJ-1 (grey) show that peak intensities are recovered when DJ-1 is present in the sample (red) but are reduced to the levels of glyc-ac-αSyn alone once DJ-1 is removed from the sample (grey). (**B**) ^{15}N-R$_2$ values of 250 μM ^{15}N-native ac-αSyn alone (black) or 250 μM ^{15}N-glyc-ac-αSyn in the presence of DJ-1 (red). The backbone dynamics of glyc-ac-αSyn with DJ-1 (red) largely overlap with native ac-αSyn without DJ-1 (black), consistent with the peak intensities in the ^{1}H–^{15}N HSQC. (**C**) Upon removal of DJ-1 (grey), the backbone dynamics as measured by ^{15}N-R$_2$ rates revert back to resembling those of glyc-ac-αSyn alone, displaying increased ^{15}N-R$_2$ in the N-terminal and NAC regions relative to native ac-αSyn (black). Error bars are determined from the fitting errors of the single exponential decay fits. All samples were monitored in PBS at 15 °C.

To further investigate the role of DJ-1 on glyc-ac-αSyn, we removed DJ-1 from the solution to determine whether the modification was permanent or due to DJ-1–glyc-ac-αSyn interactions. Strikingly, upon filtration of DJ-1 from solution, the ^{15}N-glyc-ac-αSyn ^{1}H–^{15}N HSQC spectrum reverts back to its original signature and shows peak intensities, linewidths, and chemical shifts that are indistinguishable from glyc-ac-αSyn before the addition of DJ-1 (Figure 2C). This indicates that DJ-1 does not chemically deglycate glyc-ac-αSyn, since the deglycated ac-αSyn could not be spontaneously glycated without the presence of MGO or another glycating agent. These data are supported by UV–Vis absorbance spectra, which show no substantial change in the absorbance wavelength profile between glyc-ac-αSyn in the absence of DJ-1 and after removal of DJ-1, presenting the characteristic increased absorbance from ~300–400 nm relative to native ac-αSyn (Figure S4). In addition, the reaction of DJ-1 with glyc-ac-αSyn does not enable amyloid formation, as assessed using ThT fluorescence, as would be expected if the ac-αSyn was chemically deglycated by DJ-1 (Figure S5). Together, these data suggest that DJ-1 does not effectively deglycate ac-αSyn, but rather imposes native-like structural and dynamic characteristics on glyc-ac-αSyn in solution via protein–protein interactions.

3.3. DJ-1 Interacts Primarily with Glyc-ac-αSyn Oligomers via Its Catalytic Site

In order to further characterize the specificity of the DJ-1–glyc-ac-αSyn interactions and determine a binding interface on DJ-1, we monitored ^{15}N-DJ-1 chemical shift, peak intensity, and ^{15}N-R_2 changes upon incubation with glyc-ac-αSyn using NMR spectroscopy. Upon incubating ^{15}N-DJ-1 with glyc-ac-αSyn, residue-specific ^{15}N-R_2 increases and significant line broadening are observed (Figure 3A,B). No chemical shift perturbations are observed. By mapping the residues with significant increases in ^{15}N-R_2 (Figure 3A,B, dark red) and/or line broadening (Figure 3A,B, light red) on the dimer structure of DJ-1, it is apparent that perturbations due to the presence of glyc-ac-αSyn are located primarily in the catalytic triad and surrounding residues (Figure 3C). The same interaction site on DJ-1 was also observed for interaction with native ac-αSyn (Figure S6A,B).

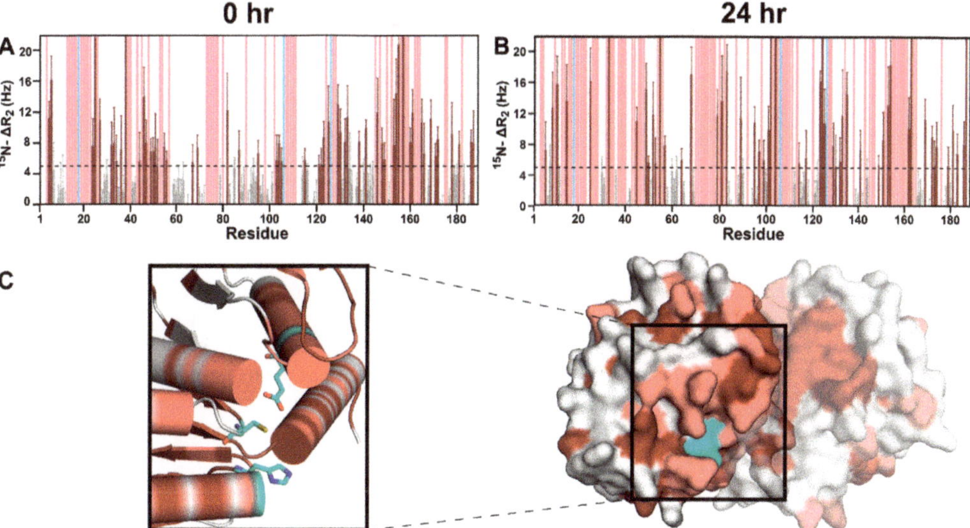

Figure 3. The catalytic site of DJ-1 is an interaction interface for glyc-ac-αSyn. (**A**) Per-residue ΔR_2 values of ^{15}N DJ-1 in the presence of glyc-ac-αSyn with 0-hour pre-incubation (no pre-incubation) showing residue-specific R_2 enhancement (dark red) or peak broadening beyond detection (light red); cyan residues indicate the catalytic triad. (**B**) ^{15}N-ΔR_2 values of ^{15}N-DJ-1 with glyc-ac-αSyn that had been pre-incubated at room temperature for 24 h to increase the concentration of oligomers in solution. The increased ^{15}N-ΔR_2 of DJ-1 in the presence of a higher concentration of glyc-ac-αSyn oligomers indicates that DJ-1 has greater binding propensity to ac-αSyn oligomers than monomers. All residues in grey are under 5 Hz change. (**C**) ^{15}N-ΔR_2 values of DJ-1 in the presence of glyc-ac-αSyn with 0-hour pre-incubation are mapped onto the 3D structure of DJ-1 (colored as in **A**), highlighting the interaction site near the catalytic triad of DJ-1. The catalytic triad is shown in cyan. All spectra were collected using a 700 MHz spectrometer at 25 °C.

To assess the necessity of the catalytic triad for the DJ-1–ac-αSyn interaction, the catalytic site C106 was either mutated to alanine or not oxidized. Upon addition of ac-αSyn to C106A-DJ-1 or non-oxidized DJ-1, the ^{15}N-R$_2$ increases seen in WT-DJ-1 are completely abolished (Figure S6C,D), suggesting that interaction of DJ-1 with ac-αSyn does not arise under these conditions. In addition, the C106A mutation or the cysteine reduction in DJ-1 dramatically suppresses ac-αSyn amyloid inhibition by DJ-1 (Figure S7). Together, these data support the role of the DJ-1 catalytic active site in the DJ-1–ac-αSyn interaction.

We probed the DJ-1 interactions with native ac-αSyn from the perspective of ac-αSyn monomers by monitoring ^{15}N-native ac-αSyn monomer NMR signals upon the addition of DJ-1. However, the co-incubation of ^{15}N-native ac-αSyn with DJ-1 in a 1:2 molar ratio resulted in no significant changes in line broadening or chemical shifts (Figure S8), despite changes in the ^{15}N-R$_2$ values of DJ-1 described above. One explanation for this is that the DJ-1 is interacting primarily with ac-αSyn oligomers rather than monomers. Thus, while we observe perturbations to DJ-1 due its interactions with native or glycated ac-αSyn oligomers, there is no observable effect on the unperturbed native ac-αSyn monomers.

To support this argument, we increased the concentration of glycated or native ac-αSyn oligomers added to ^{15}N-DJ-1 and monitored changes in DJ-1 NMR signals. Incubating glyc-ac-αSyn at room temperature for 24 h resulted in increased amounts of oligomers in the sample as observed using AFM imaging (Figure 4A,B,D). Indeed, with increased concentrations of glycated or native ac-αSyn oligomers, DJ-1 shows increased ^{15}N-R$_2$ values and peak broadening to more residues (Figure 3B and Figure S6B, compared to Figure 3A, Figure S6A), suggesting that DJ-1 primarily interacts with ac-αSyn oligomers.

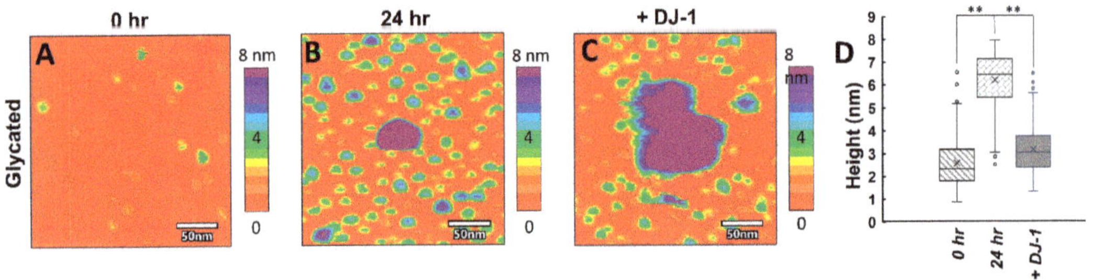

Figure 4. Liquid AFM of glyc-ac-αSyn highlighting changes in observed species upon incubation with DJ-1. (**A**) Freshly prepared glyc-ac-αSyn primarily presents as circular, compact monomers ≈3 nm in height (yellow–green). (**B**) After 24 h of incubation at room temperature, smaller oligomeric species ≈6 nm in height (green–blue) in addition to larger oligomers over 8 nm in height (purple) are apparent. (**C**) When DJ-1 is introduced to glyc-ac-αSyn that had been pre-incubated for 24 h, the aggregates coalesce, forming large, amorphous segmented clusters. However, monomers are left free on the surface, indicating that the oligomeric species are sequestered into the DJ-1 induced complexes, while monomers are left free in solution. (**D**) Distributions of heights from the above AFM samples. For this calculation only objects in the images below 8 nm in height were accounted for (0 h = 2.58 ± 1.04 nm, N = 432; 24 h = 6.21 ± 1.17 nm, N = 596; +DJ1 = 3.18 ± 1.00, N = 423). ** denotes *p*-values << 0.01.

3.4. Glyc-ac-αSyn Oligomers Participate in DJ-1 Induced Complexes, Releasing Monomers in Solution

In order to directly observe how the glyc-ac-αSyn oligomer and monomer species are perturbed upon addition of DJ-1, we used AFM in solution. Alone in solution, glyc-ac-αSyn that has been preincubated at room temperature for 24 h exists as monomeric species ~3 nm (Figure 4, green), smaller oligomers ~6–8 nm (Figure 4, blue), and larger oligomers > 8 nm (Figure 4, purple). Co-incubation of DJ-1 with preincubated glyc-ac-αSyn results in amorphous, segmented complexes (Figure 4C) that are larger than either glyc-ac-αSyn oligomers or DJ-1 in height and area. Strikingly, the addition of DJ-1 significantly reduces the number of glyc-ac-αSyn oligomers on the order of ~6–8 nm that are observed in the AFM images (Figure 4B–D), supporting their uptake into the DJ-1-induced complexes that are larger than 8 nm. Meanwhile, monomers ~3 nm in height (Figure 4, green) remain free and dispersed on the substrate, suggesting that they do not participate in the complexes. A quantitative boxplot distribution analysis of particle heights under these conditions (Figure 4D) directly demonstrates that DJ-1 interacts with oligomers while leaving the monomer glyc-ac-αSyn free in the solution. The AFM data are consistent with our NMR data in Figure 2B, which shows that upon addition of DJ-1 to glyc-ac-αSyn, the ^{15}N-R$_2$ rates of glyc-ac-αSyn revert to native-like values, representative of free ac-αSyn monomers in solution. These data support our hypothesis that glycated and native ac-αSyn oligomers interact with DJ-1, while monomers do not participate in the interactions and remain free in solution. Notably, the interaction of DJ-1 with glyc-ac-αSyn oligomers does not degrade the oligomers, as evidenced by the restoration of enhanced ^{15}N-R$_2$ rates upon removal of DJ-1 that are indistinguishable from those in the absence of DJ-1 (Figure 2C).

4. Discussion

The glycation of proteins has been shown to lead to increased protein aggregation and hindered cellular clearance and is associated with degenerative diseases such as PD, Alzheimer's disease, diabetes, and atherosclerosis. The glycation of αSyn has been shown to increase oligomer formation and produce heterogeneous amorphous aggregates and suppress amyloid formation [10,31]. Investigating the changes induced by glycation on the biophysical characteristics of αSyn can help clarify the role of glyc-αSyn in synucleinopathies. Our SEC data and AFM imaging are consistent with the literature, showing a new population of oligomers formed by glyc-ac-αSyn. Although glyc-ac-αSyn does not form amyloid fibrils, the aggregates that it does produce are toxic to cells, are unable to be cleared, and alter lipid binding to disrupt physiological function [32,67–69]. Understanding how to interfere with these toxic oligomers may aid in therapeutic design against pathological glycation.

DJ-1 is known to interact with and regulate numerous proteins implicated across various biological systems, including neurodegenerative disorders [46,70,71], diabetes [72], and cancer [73]. Numerous studies have addressed interactions of DJ-1 with various forms of native αSyn, including with monomer, oligomer, and amyloid species [49,52,55,74]. DJ-1 has been reported to have weak to minimal binding to αSyn monomers [55] and while direct interactions with αSyn oligomers have not been established in vitro [52], DJ-1 has been shown to reduce αSyn oligomerization in vivo [49,74]. In addition, DJ-1 has been shown to attenuate αSyn aggregation through chaperone-mediated autophagy [54]. Prior research shows that DJ-1 binds to aggregated forms of αSyn and that αSyn fibrils have increased toxicity following alteration by DJ-1, indicating that DJ-1 modifies aggregated forms of αSyn [53].

Here, we are interested in understanding the molecular interactions of DJ-1 with glyc-ac-αSyn monomers and oligomers. Very little is known about DJ-1 interactions with glycated proteins. However, DJ-1 has been proposed to act as a glyoxalase to help reduce the harmful effects of glycating species [33] and overexpression of DJ-1 has been shown to mitigate the glycation-induced toxicity and aggregation of αSyn [57]. Glycated αSyn monomers and oligomers can induce an increase in oxidative stress leading to further toxicity [23]. The upregulation of DJ-1 has been proven to mitigate the effects of oxidative stress by acting as a scavenger for reactive oxygen species [75–77]. Therefore, controlling DJ-1 expression is essential to reducing αSyn toxicity caused by the effects of ROS and glycation.

Based on the combination of AFM and NMR results, we propose that (1) DJ-1 interacts with glyc-ac-αSyn oligomers to sequester them into larger aggregates, (2) that the sequestration of oligomers by DJ-1 suppresses glyc-ac-αSyn monomer–oligomer interactions, and (3) that the sequestration of oligomers allows the release of free glyc-ac-αSyn monomers (Figure 5). By inhibiting these monomer–oligomer interactions, the aggregation and propagation of αSyn aggregates may be suppressed, providing an avenue for therapeutic intervention to mitigate the harmful effects of the aging-induced glycation of αSyn. We demonstrate that the catalytic triad of DJ-1, specifically C106, is largely responsible for the interaction with ac-αSyn oligomers supporting previous research on the oxidation state of DJ-1 that shows that the proper oxidation of C106 to the sulfinic acid form is imperative for function as a chaperone, its redox capability, and its cytoprotective function [38,78–80].

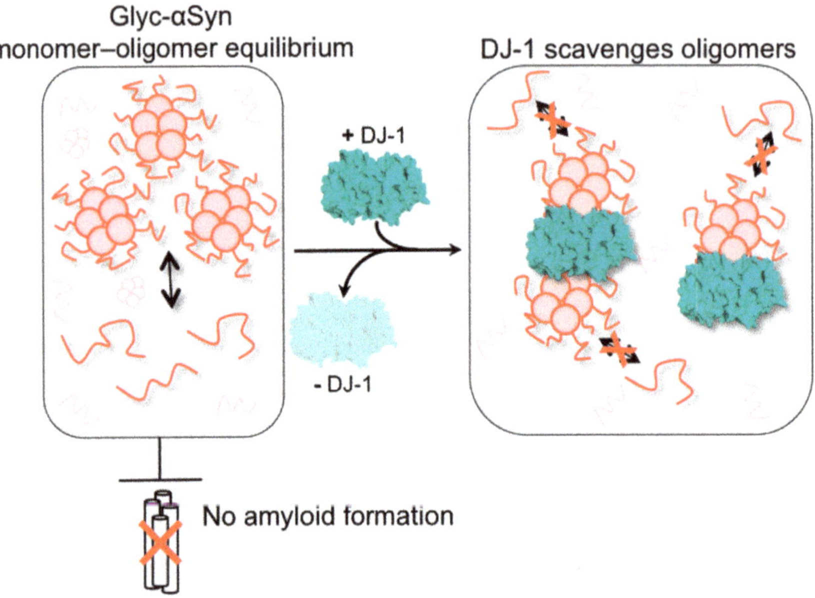

Figure 5. Proposed mechanism for DJ-1's impact on ac-αSyn glycation. (Left) Glyc-ac-αSyn spontaneously forms oligomers that are in equilibrium with glyc-ac-αSyn monomers. Glyc-ac-αSyn does not form amyloid fibrils, although the oligomers have been shown to be harmful to neurons. (Right) Upon addition of DJ-1, DJ-1 scavenges glyc-ac-αSyn oligomers, preventing their interactions with glyc-ac-αSyn monomers. This allows glyc-ac-αSyn monomers to be free in solution. Removal of DJ-1 from the system restores the glyc-ac-αSyn monomer–oligomer interactions.

DJ-1 expression and oxidation levels can potentially be biomarkers for progressive forms of Parkinson's disease [81–83]. The upregulation of DJ-1 has been proven to mitigate the effects of oxidative stress by acting as a scavenger for reactive oxygen species [75–77]. Therefore, large amounts of ROS may alter the oxidation state of DJ-1 within the cell and lead to decreased DJ-1 function as a modulator of αSyn aggregation. As a consequence of these DJ-1–glyc-ac-αSyn interactions found in our study, a decreased accumulation of glyc-ac-αSyn aggregates within dopaminergic neurons may protect against neurodegenerative effects caused by harmful αSyn aggregates. Thus, targeting the upregulation of DJ-1 and suppressing the effects caused by αSyn glycation and amyloid accumulation may aid in treatment strategies against Parkinson's disease.

Supplementary Materials: The following are available online at https://www.mdpi.com/article/10.3390/biom11101466/s1, Figure S1: Impact of DJ-1 on native αSyn amyloid formation, Figure S2: Characterization of glyc-αSyn, Figure S3: DJ-1 restores native-like features in ^{1}H–^{15}N HSQC spectra, Figure S4: Incubation with DJ-1 does not deglycate αSyn, Figure S5: Impact of αSyn glycation on amyloid formation, Figure S6: Supporting the DJ-1 catalytic site as the binding site for αSyn, Figure S7: Impact of the DJ-1 catalytic triad on αSyn amyloid formation, and Figure S8: Impact of DJ-1 on αSyn monomers. References [10,32,51–53] are also cited in the Supplementary Materials.

Author Contributions: Conceptualization, T.B.A., X.Y. and J.B.; formal analysis, T.B.A., J.R., X.Y. and C.L.H.; funding acquisition, J.B.; investigation, T.B.A., J.R. and X.Y.; project administration, J.B.; supervision, C.L.H. and J.B.; visualization, T.B.A. and J.R.; writing—original draft, T.B.A., J.R. and

C.L.H.; writing—review and editing, T.B.A., J.R., X.Y., C.L.H. and J.B. All authors have read and agreed to the published version of the manuscript.

Funding: This research was funded by the National Institutes of Health, R35 GM136431 to JB.

Institutional Review Board Statement: Not applicable.

Informed Consent Statement: Not applicable.

Data Availability Statement: Data are available upon reasonable request from the corresponding author.

Conflicts of Interest: The authors declare no conflict of interest. The funders had no role in the design of the study; in the collection, analyses, or interpretation of data; in the writing of the manuscript, or in the decision to publish the results.

References

1. Nowotny, K.; Jung, T.; Höhn, A.; Weber, D.; Grune, T. Advanced Glycation End Products and Oxidative Stress in Type 2 Diabetes Mellitus. *Biomolecules* **2015**, *5*, 194–222. [CrossRef] [PubMed]
2. Lee, A.T.; Cerami, A. Role of Glycation in Aging. *Ann. N. Y. Acad. Sci.* **1992**, *663*, 63–70. [CrossRef] [PubMed]
3. Singh, R.; Barden, A.; Mori, T.; Beilin, L. Advanced glycation end-products: A review. *Diabetologia* **2001**, *44*, 129–146. [CrossRef] [PubMed]
4. Münch, G.; Lüth, H.J.; Wong, A.; Arendt, T.; Hirsch, E.; Ravid, R.; Riederer, P. Crosslinking of α-synuclein by advanced glycation endproducts — an early pathophysiological step in Lewy body formation? *J. Chem. Neuroanat.* **2000**, *20*, 253–257. [CrossRef]
5. Sasaki, N.; Fukatsu, R.; Tsuzuki, K.; Hayashi, Y.; Yoshida, T.; Fujii, N.; Koike, T.; Wakayama, I.; Yanagihara, R.; Garruto, R.; et al. Advanced Glycation End Products in Alzheimer's Disease and Other Neurodegenerative Diseases. *Am. J. Pathol.* **1998**, *153*, 1149–1155. [CrossRef]
6. Vicente Miranda, H.; Gomes, M.A.; Branco-Santos, J.; Breda, C.; Lázaro, D.F.; Lopes, L.V.; Herrera, F.; Giorgini, F.; Outeiro, T.F. Glycation potentiates neurodegeneration in models of Huntington's disease. *Sci. Rep.* **2016**, *6*, 36798. [CrossRef]
7. Vlassara, H.; Palace, M.R. Diabetes and advanced glycation endproducts. *J. Intern. Med.* **2002**, *251*, 87–101. [CrossRef]
8. Peppa, M.; Uribarri, J.; Vlassara, H. The role of advanced glycation end products in the development of atherosclerosis. *Curr. Diabetes Rep.* **2004**, *4*, 31–36. [CrossRef]
9. Li, X.H.; Du, L.L.; Cheng, X.S.; Jiang, X.; Zhang, Y.; Lv, B.L.; Liu, R.; Wang, J.Z.; Zhou, X.W. Glycation exacerbates the neuronal toxicity of β-amyloid. *Cell Death Dis.* **2013**, *4*, e673. [CrossRef]
10. Vicente Miranda, H.; Szegő, É.M.; Oliveira, L.M.A.; Breda, C.; Darendelioglu, E.; De Oliveira, R.M.; Ferreira, D.G.; Gomes, M.A.; Rott, R.; Oliveira, M.; et al. Glycation potentiates α-synuclein-associated neurodegeneration in synucleinopathies. *Brain* **2017**, *140*, 1399–1419. [CrossRef]
11. Spillantini, M.G.; Schmidt, M.L.; Lee, V.M.-Y.; Trojanowski, J.Q.; Jakes, R.; Goedert, M. α-Synuclein in Lewy bodies. *Nature* **1997**, *388*, 839–840. [CrossRef]
12. Valdinocci, D.; Radford, R.; Siow, S.; Chung, R.; Pountney, D. Potential modes of intercellular α-synuclein transmission. *Int. J. Mol. Sci.* **2017**, *18*, 469. [CrossRef]
13. Alam, P.; Bousset, L.; Melki, R.; Otzen, D.E. α-synuclein oligomers and fibrils: A spectrum of species, a spectrum of toxicities. *J. Neurochem.* **2019**, *150*, 522–534. [CrossRef]
14. Ingelsson, M. Alpha-synuclein oligomers—Neurotoxic molecules in Parkinson's disease and other Lewy body disorders. *Front. Neurosci.* **2016**, *10*, 408. [CrossRef]
15. Kahle, P.J.; Neumann, M.; Ozmen, L.; Müller, V.; Odoy, S.; Okamoto, N.; Jacobsen, H.; Iwatsubo, T.; Trojanowski, J.Q.; Takahashi, H. Selective insolubility of α-synuclein in human Lewy body diseases is recapitulated in a transgenic mouse model. *Am. J. Pathol.* **2001**, *159*, 2215–2225. [CrossRef]
16. Sharon, R.; Bar-Joseph, I.; Frosch, M.P.; Walsh, D.M.; Hamilton, J.A.; Selkoe, D.J. The formation of highly soluble oligomers of α-synuclein is regulated by fatty acids and enhanced in Parkinson's disease. *Neuron* **2003**, *37*, 583–595. [CrossRef]
17. Periquet, M.; Fulga, T.; Myllykangas, L.; Schlossmacher, M.G.; Feany, M.B. Aggregated α-synuclein mediates dopaminergic neurotoxicity in vivo. *J. Neurosci.* **2007**, *27*, 3338–3346. [CrossRef]
18. Bengoa-Vergniory, N.; Roberts, R.F.; Wade-Martins, R.; Alegre-Abarrategui, J. Alpha-synuclein oligomers: A new hope. *Acta Neuropathol.* **2017**, *134*, 819–838. [CrossRef] [PubMed]
19. Colla, E.; Jensen, P.H.; Pletnikova, O.; Troncoso, J.C.; Glabe, C.; Lee, M.K. Accumulation of toxic α-synuclein oligomer within endoplasmic reticulum occurs in α-synucleinopathy in vivo. *J. Neurosci.* **2012**, *32*, 3301–3305. [CrossRef] [PubMed]
20. Karpinar, D.P.; Balija, M.B.G.; Kügler, S.; Opazo, F.; Rezaei-Ghaleh, N.; Wender, N.; Kim, H.Y.; Taschenberger, G.; Falkenburger, B.H.; Heise, H. Pre-fibrillar α-synuclein variants with impaired β-structure increase neurotoxicity in Parkinson's disease models. *EMBO J.* **2009**, *28*, 3256–3268. [CrossRef] [PubMed]
21. Winner, B.; Jappelli, R.; Maji, S.K.; Desplats, P.A.; Boyer, L.; Aigner, S.; Hetzer, C.; Loher, T.; Vilar, M.; Campioni, S. In vivo demonstration that α-synuclein oligomers are toxic. *Proc. Natl. Acad. Sci. USA* **2011**, *108*, 4194–4199. [CrossRef]

22. Cremades, N.; Chen, S.; Dobson, C. Structural characteristics of α-synuclein oligomers. *Int. Rev. Cell Mol. Biol.* **2017**, *329*, 79–143. [PubMed]

23. Guerrero, E.; Vasudevaraju, P.; Hegde, M.L.; Britton, G.B.; Rao, K.S. Recent Advances in α-Synuclein Functions, Advanced Glycation, and Toxicity: Implications for Parkinson's Disease. *Mol. Neurobiol.* **2013**, *47*, 525–536. [CrossRef] [PubMed]

24. Chen, L.; Wei, Y.; Wang, X.; He, R. Ribosylation Rapidly Induces α-Synuclein to Form Highly Cytotoxic Molten Globules of Advanced Glycation End Products. *PLoS ONE* **2010**, *5*, e9052. [CrossRef] [PubMed]

25. Shaikh, S.; Nicholson, L.F.B. Advanced glycation end products induce in vitro cross-linking of α-synuclein and accelerate the process of intracellular inclusion body formation. *J. Neurosci. Res.* **2008**, *86*, 2071–2082. [CrossRef] [PubMed]

26. Castellani, R.; Smith, M.A.; Richey, G.L.; Perry, G. Glycoxidation and oxidative stress in Parkinson disease and diffuse Lewy body disease. *Brain Res.* **1996**, *737*, 195–200. [CrossRef]

27. Dalfó, E.; Portero-Otín, M.; Ayala, V.; Martínez, A.; Pamplona, R.; Ferrer, I. Evidence of Oxidative Stress in the Neocortex in Incidental Lewy Body Disease. *J. Neuropathol. Exp. Neurol.* **2005**, *64*, 816–830. [CrossRef] [PubMed]

28. Thornalley, P.J.; Langborg, A.; Minhas, H.S. Formation of glyoxal, methylglyoxal and 3-deoxyglucosone in the glycation of proteins by glucose. *Biochem. J.* **1999**, *344*, 109–116. [CrossRef]

29. Ahmed, M.U.; Frye, E.B.; Degenhardt, T.P.; Thorpe, S.R.; Baynes, J.W. N ε-(Carboxyethyl)lysine, a product of the chemical modification of proteins by methylglyoxal, increases with age in human lens proteins. *Biochem. J.* **1997**, *324*, 565–570. [CrossRef]

30. Martínez-Orozco, H.; Mariño, L.; Uceda, A.B.; Ortega-Castro, J.; Vilanova, B.; Frau, J.; Adrover, M. Nitration and Glycation Diminish the α-Synuclein Role in the Formation and Scavenging of Cu2+-Catalyzed Reactive Oxygen Species. *ACS Chem. Neurosci.* **2019**, *10*, 2919–2930. [CrossRef]

31. Lee, D.; Park, C.W.; Paik, S.R.; Choi, K.Y. The modification of α-synuclein by dicarbonyl compounds inhibits its fibril-forming process. *Biochim. Biophys. Acta Proteins Proteom.* **2009**, *1794*, 421–430. [CrossRef] [PubMed]

32. Padmaraju, V.; Bhaskar, J.J.; Prasada Rao, U.J.; Salimath, P.V.; Rao, K. Role of advanced glycation on aggregation and DNA binding properties of α-synuclein. *J. Alzheimer's Dis.* **2011**, *24*, 211–221. [CrossRef] [PubMed]

33. Lee, J.-Y.; Song, J.; Kwon, K.; Jang, S.; Kim, C.; Baek, K.; Kim, J.; Park, C. Human DJ-1 and its homologs are novel glyoxalases. *Hum. Mol. Genet.* **2012**, *21*, 3215–3225. [CrossRef] [PubMed]

34. Wilson, M.A.; Collins, J.L.; Hod, Y.; Ringe, D.; Petsko, G.A. The 1.1-Å resolution crystal structure of DJ-1, the protein mutated in autosomal recessive early onset Parkinson's disease. *Proc. Natl. Acad. Sci. USA* **2003**, *100*, 9256–9261. [CrossRef] [PubMed]

35. Tao, X.; Tong, L. Crystal Structure of Human DJ-1, a Protein Associated with Early Onset Parkinson's Disease. *J. Biol. Chem.* **2003**, *278*, 31372–31379. [CrossRef] [PubMed]

36. Taira, T.; Saito, Y.; Niki, T.; Iguchi-Ariga, S.M.; Takahashi, K.; Ariga, H. DJ-1 has a role in antioxidative stress to prevent cell death. *EMBO Rep.* **2004**, *5*, 213–218. [CrossRef]

37. Kinumi, T.; Kimata, J.; Taira, T.; Ariga, H.; Niki, E. Cysteine-106 of DJ-1 is the most sensitive cysteine residue to hydrogen peroxide-mediated oxidation in vivo in human umbilical vein endothelial cells. *Biochem. Biophys. Res. Commun.* **2004**, *317*, 722–728. [CrossRef]

38. Wilson, M.A. The role of cysteine oxidation in DJ-1 function and dysfunction. *Antioxid. Redox Signal.* **2011**, *15*, 111–122. [CrossRef]

39. Junn, E.; Jang, W.H.; Zhao, X.; Jeong, B.S.; Mouradian, M.M. Mitochondrial localization of DJ-1 leads to enhanced neuroprotection. *J. Neurosci. Res.* **2009**, *87*, 123–129. [CrossRef]

40. Lev, N.; Ickowicz, D.; Barhum, Y.; Lev, S.; Melamed, E.; Offen, D. DJ-1 protects against dopamine toxicity. *J. Neural Transm.* **2009**, *116*, 151–160. [CrossRef]

41. Richarme, G.; Liu, C.; Mihoub, M.; Abdallah, J.; Leger, T.; Joly, N.; Liebart, J.-C.; Jurkunas, U.V.; Nadal, M.; Bouloc, P.; et al. Guanine glycation repair by DJ-1/Park7 and its bacterial homologs. *Science* **2017**, *357*, 208–211. [CrossRef] [PubMed]

42. Richarme, G.; Mihoub, M.; Dairou, J.; Bui, L.C.; Leger, T.; Lamouri, A. Parkinsonism-associated Protein DJ-1/Park7 Is a Major Protein Deglycase That Repairs Methylglyoxal- and Glyoxal-glycated Cysteine, Arginine, and Lysine Residues. *J. Biol. Chem.* **2015**, *290*, 1885–1897. [CrossRef] [PubMed]

43. Jun, Y.W.; Kool, E.T. Small Substrate or Large? Debate Over the Mechanism of Glycation Adduct Repair by DJ-1. *Cell Chem. Biol.* **2020**, *27*, 1117–1123. [CrossRef] [PubMed]

44. Andreeva, A.; Bekkhozhin, Z.; Omertassova, N.; Baizhumanov, T.; Yeltay, G.; Akhmetali, M.; Toibazar, D.; Utepbergenov, D. The apparent deglycase activity of DJ-1 results from the conversion of free methylglyoxal present in fast equilibrium with hemithioacetals and hemiaminals. *J. Biol. Chem.* **2019**, *294*, 18863–18872. [CrossRef] [PubMed]

45. Bonifati, V.; Rizzu, P.; van Baren, M.J.; Schaap, O.; Breedveld, G.J.; Krieger, E.; Dekker, M.C.; Squitieri, F.; Ibanez, P.; Joosse, M. Mutations in the DJ-1 gene associated with autosomal recessive early-onset parkinsonism. *Science* **2003**, *299*, 256–259. [CrossRef]

46. Antipova, D.; Bandopadhyay, R. *Expression of DJ-1 in Neurodegenerative Disorders*; Springer: Singapore, 2017; pp. 25–43.

47. Malgieri, G.; Eliezer, D. Structural effects of Parkinson's disease linked DJ-1 mutations. *Protein Sci.* **2008**, *17*, 855–868. [CrossRef]

48. Bandopadhyay, R.; Kingsbury, A.E.; Cookson, M.R.; Reid, A.R.; Evans, I.M.; Hope, A.D.; Pittman, A.M.; Lashley, T.; Canet-Aviles, R.; Miller, D.W. The expression of DJ-1 (PARK7) in normal human CNS and idiopathic Parkinson's disease. *Brain* **2004**, *127*, 420–430. [CrossRef]

49. Zondler, L.; Miller-Fleming, L.; Repici, M.; Goncalves, S.; Tenreiro, S.; Rosado-Ramos, R.; Betzer, C.; Straatman, K.; Jensen, P.H.; Giorgini, F. DJ-1 interactions with α-synuclein attenuate aggregation and cellular toxicity in models of Parkinson's disease. *Cell Death Dis.* **2014**, *5*, e1350. [CrossRef]

50. Neumann, M.; Müller, V.; Görner, K.; Kretzschmar, H.A.; Haass, C.; Kahle, P.J. Pathological properties of the Parkinson?s disease-associated protein DJ-1 in a-synucleinopathies and tauopathies: Relevance for multiple system atrophy and Pick?s disease. *Acta Neuropathol.* **2004**, *107*, 489–496. [CrossRef]

51. Zhou, W.; Zhu, M.; Wilson, M.A.; Petsko, G.A.; Fink, A.L. The oxidation state of DJ-1 regulates its chaperone activity toward α-synuclein. *J. Mol. Biol.* **2006**, *356*, 1036–1048. [CrossRef]

52. Shendelman, S.; Jonason, A.; Martinat, C.; Leete, T.; Abeliovich, A. DJ-1 is a redox-dependent molecular chaperone that inhibits α-synuclein aggregate formation. *PLoS Biol.* **2004**, *2*, e362. [CrossRef] [PubMed]

53. Kumar, R.; Kumar, S.; Hanpude, P.; Singh, A.K.; Johari, T.; Majumder, S.; Maiti, T.K. Partially oxidized DJ-1 inhibits α-synuclein nucleation and remodels mature α-synuclein fibrils in vitro. *Commun. Biol.* **2019**, *2*. [CrossRef] [PubMed]

54. Xu, C.-Y.; Kang, W.-Y.; Chen, Y.-M.; Jiang, T.-F.; Zhang, J.; Zhang, L.-N.; Ding, J.-Q.; Liu, J.; Chen, S.-D. DJ-1 inhibits α-synuclein aggregation by regulating chaperone-mediated autophagy. *Front. Aging Neurosci.* **2017**, *9*, 308. [CrossRef] [PubMed]

55. Jin, J.; Li, G.J.; Davis, J.; Zhu, D.; Wang, Y.; Pan, C.; Zhang, J. Identification of Novel Proteins Associated with Both α-Synuclein and DJ-1. *Mol. Cell. Proteom.* **2007**, *6*, 845–859. [CrossRef] [PubMed]

56. Meulener, M.C.; Graves, C.L.; Sampathu, D.M.; Armstrong-Gold, C.E.; Bonini, N.M.; Giasson, B.I. DJ-1 is present in a large molecular complex in human brain tissue and interacts with α-synuclein. *J. Neurochem.* **2005**, *93*, 1524–1532. [CrossRef]

57. Sharma, N.; Rao, S.P.; Kalivendi, S.V. The deglycase activity of DJ-1 mitigates α-synuclein glycation and aggregation in dopaminergic cells: Role of oxidative stress mediated downregulation of DJ-1 in Parkinson's disease. *Free. Radic. Biol. Med.* **2019**, *135*, 28–37. [CrossRef] [PubMed]

58. Öhrfelt, A.; Zetterberg, H.; Andersson, K.; Persson, R.; Secic, D.; Brinkmalm, G.; Wallin, A.; Mulugeta, E.; Francis, P.T.; Vanmechelen, E.; et al. Identification of Novel α-Synuclein Isoforms in Human Brain Tissue by using an Online NanoLC-ESI-FTICR-MS Method. *Neurochem. Res.* **2011**, *36*, 2029–2042. [CrossRef]

59. Anderson, J.P.; Walker, D.E.; Goldstein, J.M.; De Laat, R.; Banducci, K.; Caccavello, R.J.; Barbour, R.; Huang, J.; Kling, K.; Lee, M.; et al. Phosphorylation of Ser-129 Is the Dominant Pathological Modification of α-Synuclein in Familial and Sporadic Lewy Body Disease. *J. Biol. Chem.* **2006**, *281*, 29739–29752. [CrossRef] [PubMed]

60. Kang, L.; Janowska, M.K.; Moriarty, G.M.; Baum, J. Mechanistic Insight into the Relationship between N-Terminal Acetylation of α-Synuclein and Fibril Formation Rates by NMR and Fluorescence. *PLoS ONE* **2013**, *8*, e75018. [CrossRef] [PubMed]

61. Kang, L.; Moriarty, G.M.; Woods, L.A.; Ashcroft, A.E.; Radford, S.E.; Baum, J. N-terminal acetylation of α-synuclein induces increased transient helical propensity and decreased aggregation rates in the intrinsically disordered monomer. *Protein Sci.* **2012**, *21*, 911–917. [CrossRef] [PubMed]

62. Delaglio, F.; Grzesiek, S.; Vuister, G.W.; Zhu, G.; Pfeifer, J.; Bax, A. NMRPipe: A multidimensional spectral processing system based on UNIX pipes. *J. Biomol. NMR* **1995**, *6*, 277–293. [CrossRef] [PubMed]

63. Lee, W.; Tonelli, M.; Markley, J.L. NMRFAM-SPARKY: Enhanced software for biomolecular NMR spectroscopy. *Bioinformatics* **2015**, *31*, 1325–1327. [CrossRef] [PubMed]

64. Yang, X.; Wang, B.; Hoop, C.L.; Williams, J.K.; Baum, J. NMR unveils an N-terminal interaction interface on acetylated-α-synuclein monomers for recruitment to fibrils. *Proc. Natl. Acad. Sci. USA* **2021**, *118*, e2017452118.

65. Mihoub, M.; Abdallah, J.; Richarme, G. *Protein Repair from Glycation by Glyoxals by the DJ-1 Family Maillard Deglycases*; Springer: Singapore, 2017; pp. 133–147.

66. Matsuda, N.; Kimura, M.; Queliconi, B.B.; Kojima, W.; Mishima, M.; Takagi, K.; Koyano, F.; Yamano, K.; Mizushima, T.; Ito, Y.; et al. Parkinson's disease-related DJ-1 functions in thiol quality control against aldehyde attack in vitro. *Sci. Rep.* **2017**, *7*. [CrossRef] [PubMed]

67. Atkin, G.; Paulson, H. Ubiquitin pathways in neurodegenerative disease. *Front. Mol. Neurosci.* **2014**, *7*, 63. [CrossRef]

68. Plotegher, N.; Bubacco, L. Lysines, Achilles' heel in alpha-synuclein conversion to a deadly neuronal endotoxin. *Ageing Res. Rev.* **2016**, *26*, 62–71. [CrossRef] [PubMed]

69. Abeywardana, T.; Lin, Y.H.; Rott, R.; Engelender, S.; Pratt, M.R. Site-Specific Differences in Proteasome-Dependent Degradation of Monoubiquitinated α-Synuclein. *Chem. Biol.* **2013**, *20*, 1207–1213. [CrossRef] [PubMed]

70. Saito, Y.; Miyasaka, T.; Hatsuta, H.; Takahashi-Niki, K.; Hayashi, K.; Mita, Y.; Kusano-Arai, O.; Iwanari, H.; Ariga, H.; Hamakubo, T.; et al. Immunostaining of Oxidized DJ-1 in Human and Mouse Brains. *J. Neuropathol. Exp. Neurol.* **2014**, *73*, 714–728. [CrossRef]

71. Choi, J.; Sullards, M.C.; Olzmann, J.A.; Rees, H.D.; Weintraub, S.T.; Bostwick, D.E.; Gearing, M.; Levey, A.I.; Chin, L.-S.; Li, L. Oxidative Damage of DJ-1 Is Linked to Sporadic Parkinson and Alzheimer Diseases. *J. Biol. Chem.* **2006**, *281*, 10816–10824. [CrossRef] [PubMed]

72. Eberhard, D.; Lammert, E. *The Role of the Antioxidant Protein DJ-1 in Type 2 Diabetes Mellitus*; Springer: Singapore, 2017; pp. 173–186.

73. Kawate, T.; Tsuchiya, B.; Iwaya, K. *Expression of DJ-1 in Cancer Cells: Its Correlation with Clinical Significance*; Springer: Singapore, 2017; pp. 45–59.

74. Batelli, S.; Albani, D.; Rametta, R.; Polito, L.; Prato, F.; Pesaresi, M.; Negro, A.; Forloni, G. DJ-1 Modulates α-Synuclein Aggregation State in a Cellular Model of Oxidative Stress: Relevance for Parkinson's Disease and Involvement of HSP70. *PLoS ONE* **2008**, *3*, e1884. [CrossRef] [PubMed]

75. Andres-Mateos, E.; Perier, C.; Zhang, L.; Blanchard-Fillion, B.; Greco, T.M.; Thomas, B.; Ko, H.S.; Sasaki, M.; Ischiropoulos, H.; Przedborski, S.; et al. DJ-1 gene deletion reveals that DJ-1 is an atypical peroxiredoxin-like peroxidase. *Proc. Natl. Acad. Sci. USA* **2007**, *104*, 14807–14812. [CrossRef] [PubMed]

76. Fan, J.; Ren, H.; Jia, N.; Fei, E.; Zhou, T.; Jiang, P.; Wu, M.; Wang, G. DJ-1 Decreases Bax Expression through Repressing p53 Transcriptional Activity. *J. Biol. Chem.* **2008**, *283*, 4022–4030. [CrossRef] [PubMed]

77. Raninga, P.V.; Di Trapani, G.; Tonissen, K.F. *The Multifaceted Roles of DJ-1 as an Antioxidant*; Springer: Singapore, 2017; pp. 67–87.

78. Canet-Avilés, R.M.; Wilson, M.A.; Miller, D.W.; Ahmad, R.; McLendon, C.; Bandyopadhyay, S.; Baptista, M.J.; Ringe, D.; Petsko, G.A.; Cookson, M.R. The Parkinson's disease protein DJ-1 is neuroprotective due to cysteine-sulfinic acid-driven mitochondrial localization. *Proc. Natl. Acad. Sci. USA* **2004**, *101*, 9103–9108. [CrossRef] [PubMed]

79. Blackinton, J.; Lakshminarasimhan, M.; Thomas, K.J.; Ahmad, R.; Greggio, E.; Raza, A.S.; Cookson, M.R.; Wilson, M.A. Formation of a stabilized cysteine sulfinic acid is critical for the mitochondrial function of the parkinsonism protein DJ-1. *J. Biol. Chem.* **2009**, *284*, 6476–6485. [CrossRef] [PubMed]

80. Waak, J.; Weber, S.S.; Görner, K.; Schall, C.; Ichijo, H.; Stehle, T.; Kahle, P.J. Oxidizable residues mediating protein stability and cytoprotective interaction of DJ-1 with apoptosis signal-regulating kinase 1. *J. Biol. Chem.* **2009**, *284*, 14245–14257. [CrossRef]

81. Shi, M.; Zabetian, C.P.; Hancock, A.M.; Ginghina, C.; Hong, Z.; Yearout, D.; Chung, K.A.; Quinn, J.F.; Peskind, E.R.; Galasko, D.; et al. Significance and confounders of peripheral DJ-1 and alpha-synuclein in Parkinson's disease. *Neurosci. Lett.* **2010**, *480*, 78–82. [CrossRef] [PubMed]

82. Kahle, P.J.; Waak, J.; Gasser, T. DJ-1 and prevention of oxidative stress in Parkinson's disease and other age-related disorders. *Free. Radic. Biol. Med.* **2009**, *47*, 1354–1361. [CrossRef] [PubMed]

83. Saito, Y. *DJ-1 as a Biomarker of Parkinson's Disease*; Springer: Singapore, 2017; pp. 149–171.

 *biomolecules*

Article

Design of Inhibitors of the Intrinsically Disordered Protein NUPR1: Balance between Drug Affinity and Target Function

Bruno Rizzuti [1,2,†], Wenjun Lan [3,4,†], Patricia Santofimia-Castaño [3], Zhengwei Zhou [5], Adrián Velázquez-Campoy [2,6,7,8,9], Olga Abián [2,6,10], Ling Peng [4], José L. Neira [2,11], Yi Xia [5,*] and Juan L. Iovanna [3,*]

1 CNR-NANOTEC, SS Rende (CS), Department of Physics, University of Calabria, Via P. Bucci, Cubo 31 C, 87036 Rende, Cosenza, Italy; bruno.rizzuti@cnr.it

2 Instituto de Biocomputación y Física de Sistemas Complejos, Joint Units IQFR-CSIC-BIFI, and GBsC-CSIC-BIFI, Universidad de Zaragoza, 50018 Zaragoza, Spain; adrianvc@unizar.es (A.V.-C.); oabifra@unizar.es (O.A.); jlneira@umh.es (J.L.N.)

3 Centre de Recherche en Cancérologie de Marseille (CRCM), INSERM U1068, CNRS UMR 7258, Institut Paoli-Calmettes, Aix-Marseille Université, 13288 Marseille, France; wenjun.lan@etu.univ-amu.fr (W.L.); patricia.santofimia@inserm.fr (P.S.-C.)

4 Aix-Marseille Université, CNRS, Centre Interdisciplinaire de Nanoscience de Marseille, UMR 7325, «Equipe Labellisée Ligue Contre le Cancer», 13288 Marseille, France; ling.peng@univ-amu.fr

5 Chongqing Key Laboratory of Natural Product Synthesis and Drug Research, School of Pharmaceutical Sciences, Chongqing University, Chongqing 401331, China; zhouzhengwei610@126.com

6 Aragon Institute for Health Research (IIS Aragon), 50009 Zaragoza, Spain

7 Centro de Investigación Biomédica en Red en el Área Temática de Enfermedades Hepáticas y Digestivas (CIBERehd), 28029 Barcelona, Spain

8 Departamento de Bioquímica y Biología Molecular y Celular, Universidad de Zaragoza, 50009 Zaragoza, Spain

9 Fundacion ARAID, Government of Aragon, 50018 Zaragoza, Spain

10 Instituto Aragonés de Ciencias de la Salud (IACS), 50009 Zaragoza, Spain

11 IDIBE, Universidad Miguel Hernández, 03202 Elche, Alicante, Spain

* Correspondence: yixia@cqu.edu.cn (Y.X.); juan.iovanna@inserm.fr (J.L.I.)

† These authors contributed equally to this paper.

Citation: Rizzuti, B.; Lan, W.; Santofimia-Castaño, P.; Zhou, Z.; Velázquez-Campoy, A.; Abián, O.; Peng, L.; Neira, J.L.; Xia, Y.; Iovanna, J.L. Design of Inhibitors of the Intrinsically Disordered Protein NUPR1: Balance between Drug Affinity and Target Function. *Biomolecules* **2021**, *11*, 1453. https://doi.org/10.3390/biom11101453

Academic Editor: Simona Maria Monti

Received: 26 August 2021
Accepted: 28 September 2021
Published: 3 October 2021

Publisher's Note: MDPI stays neutral with regard to jurisdictional claims in published maps and institutional affiliations.

Abstract: Intrinsically disordered proteins (IDPs) are emerging as attractive drug targets by virtue of their physiological ubiquity and their prevalence in various diseases, including cancer. NUPR1 is an IDP that localizes throughout the whole cell, and is involved in the development and progression of several tumors. We have previously repurposed trifluoperazine (TFP) as a drug targeting NUPR1 and, by using a ligand-based approach, designed the drug ZZW-115 starting from the TFP scaffold. Such derivative compound hinders the development of pancreatic ductal adenocarcinoma (PDAC) in mice, by hampering nuclear translocation of NUPR1. Aiming to further improve the activity of ZZW-115, here we have used an indirect drug design approach to modify its chemical features, by changing the substituent attached to the piperazine ring. As a result, we have synthesized a series of compounds based on the same chemical scaffold. Isothermal titration calorimetry (ITC) showed that, with the exception of the compound preserving the same chemical moiety at the end of the alkyl chain as ZZW-115, an increase of the length by a single methylene group (i.e., ethyl to propyl) significantly decreased the affinity towards NUPR1 measured in vitro, whereas maintaining the same length of the alkyl chain and adding heterocycles favored the binding affinity. However, small improvements of the compound affinity towards NUPR1, as measured by ITC, did not result in a corresponding improvement in their inhibitory properties and in cellulo functions, as proved by measuring three different biological effects: hindrance of the nuclear translocation of the protein, sensitization of cells against DNA damage mediated by NUPR1, and prevention of cancer cell growth. Our findings suggest that a delicate compromise between favoring ligand affinity and controlling protein function may be required to successfully design drugs against NUPR1, and likely other IDPs.

Keywords: intrinsically disordered proteins; nuclear protein 1; drug discovery; ligand-based design; isothermal titration calorimetry; biological assays

1. Introduction

Intrinsically disordered proteins (IDPs) have very high flexibility, hence do not maintain a stable conformation. IDPs are frequently involved in regulatory and signaling processes and their functions include cell cycle control, transcriptional regulation, replication, differentiation, and RNA processing, which are mediated by protein-protein and protein-nucleic acid interactions [1–4], as well as the formation of membraneless organelles via liquid-liquid phase separation [5]. Furthermore, IDPs are engaged in multivalent and/or promiscuous interactions, and, in fact, proteins involved in several biological pathways have a large proportion of disorder in their sequences [6].

NUPR1 (UniProtKB O60356), or nuclear protein 1, is an 82-residue-long (8 kDa), highly basic IDP, whose proper function is unknown, although it was first described as being activated in the exocrine pancreas in response to the cellular injury induced by pancreatitis [7]. NUPR1 is involved in cell-cycle regulation; in fact, the inducible expression of the *Nupr1* gene has been observed under several stress conditions [8,9]. NUPR1 is over-expressed in almost any, if not all, cancer tissues [8,10,11]. Moreover, it is involved in several protein cascades, such as in the processes regulating apoptosis through interaction with the oncoprotein ProTα [12], DNA repair in combination with the male specific lethal protein 1 [13,14], and cell-differentiation processes together with Polycomb proteins [15]. We have previously shown that genetic inactivation of *Nupr1* antagonizes the growth of pancreatic cancer [16,17]; furthermore, its inactivation stops the growth of hepatocarcinoma [18], cholangiocarcinoma [19], non-small cell lung cancer [20], glioblastoma [21], osteosarcoma [22] and multiple myeloma [23]. Collectively, these results indicate that NUPR1 could be exploited as a target to develop new therapies against cancer, by hindering its protein-protein interactions (PPIs) with other molecular partners. However, NUPR1 as a target for drug discovery is challenging due to its unfolded and dynamic nature.

We have recently developed a combination of biochemical, biophysical, bioinformatics, and biological approaches for an in vitro, in vivo, in silico, and in cellulo molecular screening, to select potential drug candidates against NUPR1, following a bottom-up approach and a lead-based strategy [24]. We have first repurposed the antipsychotic agent trifluoperazine (TFP) as an anticancer drug candidate, because it completely stops the growth of tumor in human pancreatic cancer xenograft mice [25]. However, TFP also causes neurological effects on treated mice at the doses necessary to obtain its antitumor action, precluding its use to treat cancer in clinics. To overcome these issues and increase TFP anticancer activity, we have explored ligand-based drug design (LBDD) with a rational in silico approach, modifying TFP by adding different functional groups [26]. As a result, we have identified a lead compound, ZZW-115, which bears the substitution of the methyl group in one of the nitrogen atoms of the piperazine group of TFP by an alkylated amine group (Figure 1). ZZW-115 has a far better antitumor activity than TFP and no appreciable side effects in mice. It induces cell death by both necroptotic and apoptotic mechanisms, with a mitochondrial metabolism failure that triggers lower production of ATP and overproduction of reactive oxygen species [26]. We have also demonstrated that ZZW-115 hampers NUPR1 nuclear translocation [27].

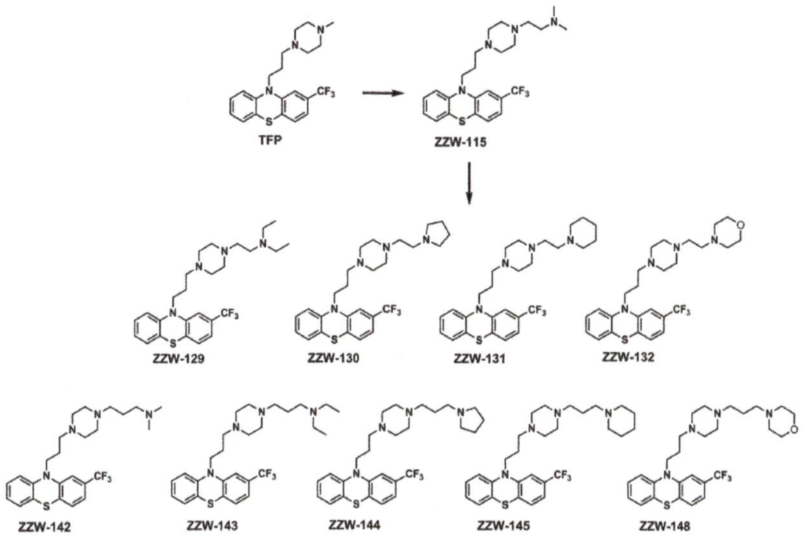

Figure 1. Structures of the ZZW-115-derived compounds used in this work. The structures of TFP and ZZW-115 are also shown for comparison.

The success of identifying the lead compound ZZW-115 against NUPR1 has stimulated us to propose further improvements in the anticancer activity of ZZW-115 by using a similar combined approach as previously pursued. In particular, our molecular optimization strategy falls within the realm of LBDD, also known as indirect drug design [28,29], which is based on the evolution of the molecular structure of compounds with known binding properties against the protein target, despite the absence of the knowledge of a binding site. In principle, even an indirect drug design should completely rely on a rational approach, possibly guided by the identification of a pharmacophore model that summarizes the key intermolecular interactions driving the binding to the specific biological target [30]. However, this is particularly challenging in the case of IDPs, due to subtle dynamic effects that govern the binding of ligands to disordered regions. For this reason, LBDD can be supplemented with a more empirical approach which consists in adding functional groups using synthetic chemical methodologies. Thus, we have used a combination of the two approaches to optimize ZZW-115 with the aim of improving its affinity towards NUPR1 and, then, its anti-cancer activity.

As a result of our strategy, in this work, we have designed and synthesized nine derivatives that maintain the basic scaffold of ZZW-115, but contain larger aromatic or alkylated groups at the distal nitrogen of the piperazine ring (Figure 1). Biophysical experiments in vitro on NUPR1, by using isothermal titration calorimetry (ITC), showed that the affinity of these derived compounds was substituent-dependent and, as a general trend, their binding energy towards NUPR1 varied with the size of the introduced chemical group. However, results obtained from cell experiments—namely, inhibition of nuclear translocation of NUPR1, sensitization of cells against DNA-damage mediated by it, and hindrance of cancer cell growth—indicate that a more favorable affinity in the binding does not necessarily correlate with the biological effects. At the end, the biological effect of a given compound depends on the binding affinity to the primary target, the effective compound concentration at the appropriate location, and the binding affinity to any alternative competing targets present in the complex cell environment. Thus, in designing compounds targeting this IDP, there must be a subtle compromise between increasing drug affinity and altering the protein function, with other properties, such as solubility, crowding, membrane permeation, cellular efflux and cellular metabolism, possibly playing an additional relevant role.

2. Materials and Methods

2.1. Materials

Ampicillin and isopropyl-β-D-1-tiogalactopyranoside were obtained from Apollo Scientific (Stockport, UK). Imidazole, Trizma base and His-Select HF nickel resin were from Sigma-Aldrich (Madrid, Spain). Triton X-100, and protein marker (PAGEmark Tricolor) were from VWR (Barcelona, Spain). The compound 5-fluorouracile (5-FU) was offered by Institute Paoli-Calmettes. Amicon centrifugal devices with a cut-off molecular weight of 3 kDa were from Millipore (Barcelona, Spain). The rest of the materials were of analytical grade. Water was deionized and purified on a Millipore system. Piperazine analogs, 2-(trifluoromethyl)phenothiazine and other reagents were purchased from Adamas-beta (Shanghai, China) or Energy Chemical (Shanghai, China).

2.2. Chemistry: General Methods

All compounds were purified by performing flash chromatography on silica gel (200–300 mesh) or preparative thin-layer chromatography. ^{1}H-NMR and ^{13}C-NMR spectra were recorded on Agilent DD2 400-MR. The chemical shifts were recorded in parts per million (ppm) with tetramethylsilane as the internal reference. The electrospray ionization mass spectroscopy (ESI-MS) data were acquired on a Waters Acquity SQ Detector mass spectrometer or Finigan LCQ mass spectrometer. The high-resolution spectra of ESI-MS were recorded on Bruker SolariX 7.0 T mass spectrometer or IonSpec 4.7 Tesla Fourier Transform mass spectrometer. All MS analysis samples were prepared as solutions in methanol. Analytical HPLC runs were performed on Waters 1525 using columns packed with Inertsil® ODS-3 5mm 4.6 × 250 mm (method 1) and Inertsil® hypersil C8 4.6 × 250 mm and (method 2) produced by GL sciences Inc, with 2489 UV/Visible detector, wavelength 254 nm. All the samples were dissolved in methanol (MeOH). The mobile phase consisted of an isocratic elution of CH_3CN/CH_3OH (30/70) with 0.1% TFA (trifluoracetic acid). The flow rate was 1.0 mL/min. Temperature was 25 °C.

2.3. Protein Expression and Purification

Expression and purification of NUPR1 was carried out as described earlier by using Ni-affinity and gel filtration chromatography [12–15,27]. The purity of NUPR1 was in all cases larger than 95%, as judged by visual inspection in SDS-PAGE. Protein concentration was determined from the absorbance of the two tyrosines in the sequence of NUPR1 [31]. We have previously shown [25,26] that aggregation of the protein does not occur in the presence of ZZW-115 or its other derivatives; only the binding of NUPR1 to the corresponding compound was observed.

2.4. Computational Modelling of ZZW-115-Derived Compounds

The binding of ZZW-115 and derived compounds to NUPR1 was modeled in silico following a procedure previously described [26]. In brief, capped fragments of the protein structure with a length of seven amino acid residues and belonging to the hot-spots of NUPR1 (regions around Ala33 and Thr68 residues) were used as a host for molecular docking of the compounds, performed by using the software AutoDock Vina [32]. The small molecular complexes obtained, mimicking each compound bound to a transient protein binding pocket, were refined in 1 ns molecular dynamics (MD) simulations carried out in the isobaric-isothermal ensemble using the GROMACS simulation package [33]. The force fields used were AMBER ff99SB-ILDN [34] for the protein and GAFF [35] for the compounds, whereas full hydration in explicit solvent was obtained by using the IDP-specific water model TIP4P-D [36]. Other simulation conditions (including modelling of electrostatics and van der Waals interactions, use of thermostat and barostat, and treatment of periodic boundary conditions) were as previously described [37,38]. After equilibration in MD simulation runs, the binding affinity of the compounds was re-evaluated using AutoDock Vina in score-only mode [32].

2.5. Organic Synthesis of ZZW-115-Derived Compounds

The ZZW-115-derived compounds were synthesized as described previously [26]. To a solution of compound **a** (50 mg, 0.13 mmol) in dimethyl-formamide (DMF) (3 mL), the corresponding piperazine derivatives (0.26 mmol) were added. The mixture was stirred at room temperature overnight under nitrogen atmosphere in dark. Then, the reaction system was concentrated and the desired product was isolated by column chromatography (CH_2Cl_2: MeOH: NH_4OH = 20:1:0.1). The analytical data of the synthetic ZZW-115 derivatives are presented below:

ZZW-129. Colorless wax. HPLC: t_R = 5.271 min (Method 1, purity 98.8%), t_R = 4.603 min (Method 2, purity > 99%). ^{1}H-NMR (400 MHz, $CDCl_3$): δ 7.04–7.19 (m, 4H, phenyl-H), 6.97 (s, 1H, phenyl-H), 6.84–6.89 (m, 2H, phenyl-H), 3.88 (t, 2H, *J* = 6.8 Hz, -CH_2-), 2.36–2.51 (m, 16H, -CH_2-), 1.83-1.90 (m, 4H, -CH_2-), 0.96 (t, 6H, *J* = 7.2 Hz, -CH_3). ^{13}C-NMR (100 MHz, $CDCl_3$): δ 145.68, 144.28, 129.87, 129.37, 127, 60, 127.51, 127.40, 124.00, 123.41, 118.99, 115.89, 111.89, 56.33, 55.40, 53.65, 53.18, 49.96, 47.32, 45.31, 29.69. MS (ESI, *m*/*z*): 492.26 [M + H]$^+$. HRMS: calcd for $C_{26}H_{35}F_3N_4S$, [M + H]$^+$, 493.2535; found, 493.2619.

ZZW-130. Colorless wax. HPLC: t_R = 5.321 min (Method 1, purity > 99%), t_R = 4.637 min (Method 2, purity > 99%). ^{1}H-NMR (400 MHz, $CDCl_3$): δ 7.10–7.19 (m, 4H, phenyl-H), 7.03 (s, 1H, phenyl-H), 6.90–6.96 (m, 2H, phenyl-H), 3.95 (t, 2H, *J* = 6.8 Hz, -CH_2-), 2.45–2.62 (m, 18H, -CH_2-), 1.90–1.97 (m, 2H, -CH_2-), 1.77 (s, 4H, -CH_2-). ^{13}C-NMR (100 MHz, $CDCl_3$): δ 145.64, 144.83, 137.41, 127.48, 127.17, 125.09, 122.06, 120.64, 115.70, 114.61, 114.53, 56.82, 56.62, 53.60, 53.16, 45.93, 45.81, 45.28, 24.33, 16.53, 16.39. MS (ESI, *m*/*z*): 491.53 [M + H]$^+$. HRMS: calcd for $C_{26}H_{33}F_3N_4S$, [M + H]$^+$, 491.2378; found, 443.2443.

ZZW-131. Colorless wax. HPLC: t_R = 5.907 min (Method 1, purity > 99%), t_R = 4.444 min (Method 2, purity > 99%). ^{1}H-NMR (400 MHz, $CDCl_3$): δ 7.09–7.18 (m, 4H, phenyl-H), 7.03 (s, 1H, phenyl-H), 6.90–6.94 (m, 2H, phenyl-H), 3.94 (t, 2H, *J* = 8.4 Hz, -CH_2-), 2.40–2.51 (m, 18H, -CH_2-), 1.90–1.95 (m, 2H, -CH_2-), 1.55–1.59 (m, 4H, -CH_2-), 1.42 (s, 2H, -CH_2-). ^{13}C-NMR (100 MHz, $CDCl_3$): δ 145.65, 144.26, 129.83, 129.62, 129.41, 127.55, 127.45, 127.34, 123.96, 123.22, 122.99, 118.90, 115.86, 111.82, 56.60, 55.96, 55.38, 55.03, 53.62, 53.23, 45.31, 25.92, 24.30, 24.12. MS (ESI, *m*/*z*): 505.54 [M + H]$^+$. HRMS: calcd for $C_{27}H_{35}F_3N_4S$, [M + H]$^+$, 505.2535; found, 505.2599.

ZZW-132. Colorless wax. HPLC: t_R = 4.844 min (Method 1, purity 98.9%), t_R = 4.218 min (Method 2, purity > 99%). ^{1}H-NMR (400 MHz, $CDCl_3$): δ 7.09–7.18 (m, 4H, phenyl-H), 7.03 (s, 1H, phenyl-H), 6.89-6.95 (m, 2H, phenyl-H), 3.94 (t, 2H, *J* = 6.8 Hz, -CH_2-), 3.69 (t, 4H, *J* = 4.4 Hz, -CH_2-), 2.45-2.49 (m, 18H, -CH_2-), 1.89-1.96 (m, 2H, -CH_2-). ^{13}C-NMR (100 MHz, $CDCl_3$): δ 145.63, 144.22, 129.82, 129.64, 129.32, 127.53, 127.44, 127.33, 125.46, 123.94, 122.98, 118.91, 118.87, 115.83, 111.83, 111.79, 66.87, 56.28, 55.53, 55.30, 54.07, 53.57, 53.15, 45.25, 24.07. MS (ESI, *m*/*z*): 507.51 [M + H]$^+$. HRMS: calcd for $C_{26}H_{33}F_3N_4S$, [M + H]$^+$, 507.2327; found, 507.2399.

ZZW-142. Colorless wax. HPLC: t_R = 5.667 min (Method 1, purity > 99%), t_R = 4.936 min (Method 2, purity > 99%). ^{1}H-NMR (400 MHz, $CDCl_3$): δ 7.20–7.10 (m, 4H, phenyl-H), 7.03 (s, 1H, phenyl-H), 6.96–6.91 (m, 2H, phenyl-H), 3.95 (t, 2H, *J* = 7.0 Hz, -CH_2-), 2.49–2.25 (m, 14H, -CH_2-), 2.21 (s, 6H, -CH_3), 1.97–1.90 (m, 2H, -CH_2-), 1.69–1.61 (m, 2H, -CH_2-). ^{13}C-NMR (100 MHz, $CDCl_3$): δ 145.64, 144.21, 129.86, 129.64, 127.58, 127.48, 127.38, 123.97, 123.03, 118.96. 118.92, 115.87, 111.85, 111.82, 57.37, 56.06, 55.28, 53.04, 53.01, 45.23, 44.78, 29.66, 29.28, 24.07, 24.03. MS (ESI, *m*/*z*): 478.24 [M + H]$^+$. HRMS: calcd for $C_{25}H_{33}F_3N_4S$, [M + H]$^+$, 479.2378; found, 479.2475.

ZZW-143. Colorless wax. HPLC: t_R = 5.264 min (Method 1, purity > 99%), t_R = 4.371 min (Method 2, purity 96.6%). ^{1}H-NMR (400 MHz, $CDCl_3$): δ 6.92–7.20 (m, 7H, phenyl-H), 3.87 (t, 2H, *J* = 6.8 Hz, -CH_2-), 2.22–2.50 (m, 18H, -CH_2-), 1.84–1.89 (m, 2H, -CH_2-), 1.53–1.61 (m, 2H, -CH_2-), 0.95 (t, 6H, *J* = 6.8 Hz, -CH_3). ^{13}C-NMR (100 MHz, $CDCl_3$): δ 145.69, 144.29, 129.88, 129.39, 127.58, 127.48, 127.37, 124.01, 123.02, 118.96, 115.89, 111.88,

111.84, 56.68, 55.41, 53.27, 53.22, 50.74, 46.80, 45.36, 29.66, 24.16, 11.34. MS (ESI, m/z): 507.26 [M + H]$^+$. HRMS: calcd for $C_{27}H_{37}F_3N_4S$, [M + H]$^+$, 507.2691; found, 507.2766.

ZZW-144. Colorless wax. HPLC: t_R = 5.314 min (Method 1, purity 98.9%), t_R = 4.641 min (Method 2, purity 97.5%). ^{1}H-NMR (400 MHz, CDCl$_3$): δ 6.90–7.20 (m, 7H, phenyl-H), 3.95 (t, 2H, J = 6.8 Hz, -CH$_2$-), 2.80–2.96 (m, 4H, -CH$_2$-), 2.36–2.49 (m, 8H, -CH$_2$-), 1.81–1.96 (m, 18H, -CH$_2$-). ^{13}C-NMR (100 MHz, CDCl$_3$): δ 145.69, 144.30, 129.88, 129.72, 129.40, 127.58, 127.49, 127.37, 124.01, 123.02, 118.96, 118.93, 115.89, 111.86, 111.85, 56.81, 55.42, 54.65, 54.20, 53.28, 53.23, 45.37, 32.63, 29.67, 26.48, 24.17, 23.39. MS (ESI, m/z): 505.24 [M + H]$^+$. HRMS: calcd for $C_{27}H_{35}F_3N_4S$, [M + H]$^+$, 505.2535; found, 505.2598.

ZZW-145. Colorless wax. HPLC: t_R = 5.088 min (Method 1, purity 97.2%), t_R = 4.434 min (Method 2, purity 97.2%). ^{1}H-NMR (400 MHz, CDCl$_3$): δ 6.90–7.20 (m, 7H, phenyl-H), 3.95 (t, 2H, J = 6.8 Hz, -CH$_2$-), 2.30–2.49 (m, 14H, -CH$_2$-), 1.92–1.97 (m, 6H, -CH$_2$-), 1.70–1.74 (m, 2H, -CH$_2$-), 1.60–1.65 (m, 4H, -CH$_2$-), 1.43–1.47 (m, 2H, -CH$_2$-). ^{13}C-NMR (100 MHz, CDCl$_3$): δ 145.67, 144.27, 129.85, 129.67, 129.37, 127.56, 127.46, 127.34, 125.48, 123.98, 122.99, 122.78, 118.93, 118.89, 115.87, 111.86, 111.82, 57.34, 56.71, 55.37, 54.50, 53.24, 53.17, 45.33, 29.65, 25.78, 24.32, 24.20, 24.14. MS (ESI, m/z): 519.51 [M + H]$^+$. HRMS: calcd for $C_{28}H_{37}F_3N_4S$, [M + H]$^+$, 519.2691; found, 519.2763.

ZZW-148. Colorless wax. HPLC: t_R = 4.841 min (Method 1, purity 98.6%), t_R = 4.897 min (Method 2, purity > 99%). ^{1}H-NMR (400 MHz, CDCl$_3$): δ 7.01–7.10 (m, 4H, phenyl-H), 6.95 (s, 1H, phenyl-H), 6.82–6.87 (m, 2H, phenyl-H), 3.87 (t, 2H, J = 6.8 Hz, -CH$_2$-), 3.62 (d, 4H, J = 4.8 Hz, -CH$_2$-), 2.26–2.41 (m, 18H, -CH$_2$-), 1.82–1.88 (m, 2H, -CH$_2$-), 1.56–1.63 (m, 2H, -CH$_2$-). ^{13}C-NMR (100 MHz, CDCl$_3$): δ 145.64, 144.21, 129.84, 127.54, 127.44, 127.33, 123.96, 122.99, 118.91, 118.87, 115.84, 111.84, 111.80, 66.88, 56.97, 56.47, 53.27, 53.67, 53.12, 53.08, 45.26, 29.62, 24.07, 23.85. MS (ESI, m/z): 521.66 [M + H]$^+$. HRMS: calcd for $C_{27}H_{35}F_3N_4OS$, [M + H]$^+$, 521.2484; found, 521.2564.

2.6. Cell Production and Viability Assays

Primary pancreatic cancer cells PDAC001T, PDAC012T, PDAC021T, PDAC081T, PDAC082T, PDAC087T, PDAC088T, PDAC089T, and PDAC115T were obtained from the PaCaOmics clinical trial registered at www.clinicaltrials.gov (accessed on 2 October 2021) with registration number NCT01692873. We included PDAC samples from both echoendoscopic ultrasound-guided fine-needle (EUS-FNA) biopsies for patients with unresectable tumors and from surgical specimens for patients undergoing surgery. The tumor samples of these patients were used to generate patient-derived xenograft in nude mice as previously reported [39]. Xenografts obtained from mice were split into several small pieces and used for cell culture in a biosafety chamber: after fine mincing, they were treated with collagenase type V (ref C9263; Sigma-Aldrich, St. Louis, MO, USA) and trypsin/EDTA (ref 25200-056; Gibco, Life Technologies, Grand Island, NY, USA) and were suspended in Dulbecco's modified Eagle's medium supplemented with 1% w/w penicillin/streptomycin (Gibco, Life Technologies, Paisley, UK) and 10% fetal bovine serum (Lonza Inc., Walkersville, MD, USA). After centrifugation, cells were resuspended in serum-free ductal media adapted from Schreiber et al. [40] at 37 °C in a 5% CO$_2$ incubator. Amplified cells were stored in liquid nitrogen as previously reported [41]. Cells were weaned from antibiotics for $\geq$48 h before testing.

Cell viability was evaluated in ten pancreatic cancer cell lines, including MiaPaCa-2, and nine primary pancreatic cancer-derived cell lines: PDAC001T, PDAC012T, PDAC021T, PDAC081T, PDAC082T, PDAC087T, PDAC088T, PDAC089T, and PDAC115T. MiaPaCa-2 cells were obtained from ATCC (Manassas, VA, USA) and maintained in DMEM (Invitrogen, Cergy-Pontoise, France), supplemented with 10% phosphate buffer solution (PBS) at 37 °C with 5% CO$_2$. The primary pancreatic cancer-derived cells were cultured in serum free ductal media (SFDM) following a procedure adapted from those previously described [40], without antibiotic, and incubated at 37 °C in a 5% CO$_2$ incubator. Cells were plated in 96-well plates (5000 cells/well) overnight. Then, the media were supplemented with the compounds to be tested at 0–100 µM concentration, and the samples were incubated for another additional 72 h before performing the measurement. Cell viability was estimated after addition of PrestoBlue™ reagent (Life Technologies, Paris, France) for 3 h according to the PrestoBlue™ cell viability reagent protocol provided by the supplier. Cell viability was normalized when comparing to untreated cell rates. Experiments were performed in triplicate and each set was repeated three times, with similar results.

2.7. Isothermal Titration Calorimetry (ITC)

The binding of the ZZW-115-derived compounds to NUPR1 was determined as described [26] by using a high sensitivity isothermal titration calorimeter Auto-iTC200 (MicroCal, Malvern-Panalytical, Malvern, UK). Protein samples and solutions were properly degassed. Experiments were performed with freshly prepared protein solutions at 25 °C. A solution of NUPR1 (20 µM, in sodium phosphate 20 mM, pH 7.0, 2% DMSO) in the calorimetric cell was titrated with a solution of each compound (200 µM, in sodium phosphate 20 mM pH 7.0, 2% DMSO). The inhibitors were dissolved in the same buffer used for the protein. A standard protocol was employed: 19 titrant injections with 2 µL compound solution were programmed with a time spacing of 150 s, a stirring speed of 750 rpm, and a reference power of 10 µcal/s. The heat evolved after each ligand injection was calculated from the integral of the calorimetric signal. The heat due to the binding reaction was obtained as the difference between the reaction heat and the corresponding heat of injection, the latter estimated as a constant value throughout the experiment, and included as an adjustable parameter in the analysis. Control experiments (with the compounds injected into the buffer) were performed under the same experimental conditions.

The association constant (K_a), the enthalpy change (ΔH) and the stoichiometry (n) of the binding reaction were obtained through nonlinear regression analysis of experimental data to a model assuming a single ligand binding site for the protein. Experiments were performed in replicate and data were analyzed using in-house-developed software implemented in Origin 7 (OriginLab).

2.8. Immunofluorescence Staining

Cells (100000 cells/well) were seeded in 12-well plates on coverslips overnight and then treated with 3 µM concentration of each compound for 24 h. After fixation with 4% paraformaldehyde (Sigma-Aldrich, St. Quentin, Fallavier, France) for 15 min, cells were washed twice with 1x PBS and incubated 1 h with the following antibodies at 1:200 dilution: rabbit anti-NUPR1 primary antibody (homemade) or γH2AX primary antibody (ab26350, Abcam, Paris, France). After the washing steps, samples were incubated 1 h with secondary antibodies at 1:500 dilution (goat anti-rabbit Alexa Fluor 488, A27034, or goat anti-mouse Alexa Fluor 488, A28175, both from Thermo Fisher Scientific, Franklin, MA, USA). ProLong™ Gold Antifade Mountant with DAPI (P36931, Thermo Fisher Scientific) was used to stain the nucleus and seal the samples. Image acquisition of Alexa Fluor 488-derived fluorescence and DAPI staining was detected with an LSM 880 controlled by Zeiss Zen Black 63x lens. Co-localization analysis and measurements in both channels were performed by using the ImageJ Coloc 2 plugin (version 3.0.5).

2.9. Genotoxicity Evaluation for Compounds Combination

The 5-FU-triggered DNA damage enhanced by ZZW-115-derived compounds was evaluated by counting the number of γH2AX foci present in each cell nucleus after immunofluorescence staining. MiaPaCa-2 cells were treated with 5-FU (10 µM), alone or in combination with the corresponding ZZW-115-derived compounds (1.5 µM), and the DNA damages caused were quantified after 12 h incubation. Untreated cells were used as a negative control.

2.10. Statistics

Statistical analyses were performed by using the unpaired two-tailed Student t test or one-way ANOVA with Tukey's *post hoc* test. Values are expressed as mean $\pm$ SEM (standard deviation of measurements). Data are representative of at least three independent experiments with triplicates completed. A value of P lower than 0.05 was considered significant.

3. Results

3.1. Docking of Compounds to a Simulated Binding Pocket of NUPR1

A computational modelling of ZZW-115-derived compounds was carried out by using a ligand-based approach, following the same protocol we have used in our previous endeavor leading to the development of ZZW-115 and other related analogs [26]. Molecular docking can reveal the anchoring of drugs to the binding hot spots of NUPR1 in a blind search on the whole simulated protein structure, as already demonstrated for TFP [25]. However, to reduce the conformational search in the identification of new compounds, it is convenient to consider fragments encompassing seven amino acid residues centered around the key protein residues Ala33 and Thr68, as previously reported [26]. These regions have been described to be at the center of the hot spots of the proteins, on the basis of two main pieces of evidence: mutational studies, and in silico screening of different compounds [15,26]. These residues also belong to the most hydrophobic regions of the protein [25]. This methodology is supported by the observation that short-range interactions, and especially local hydrophobicity, are crucial to model the binding in NUPR1 [25,26,42].

Small modifications in the molecular structure of ZZW-115 were tested by inserting the new compound within the protein fragments by molecular docking, and considering the most favorable binding modes obtained. These small complexes, each mimicking a different conformation of a transient binding pocket, were further equilibrated in MD simulations. At the end, the binding energy of the compound was obtained by evaluating it with the scoring function of AutoDock Vina [32], and averaged on the various conformations. The calculated binding scores were compared with the one obtained for ZZW-115, −7.5 ± 0.3 kcal/mol, which is largely due to the molecular scaffold of TFP (–7.0 kcal/mol) [25].

The results showed that variations in the propyl linker between the phenothiazine and piperazine ring did not increase the binding affinity. In contrast, addition of a methylene group in the linker between the piperazine ring and the alkylated amine group (leading to compound ZZW-142; see Figure 1) improved the average binding score by –0.3 kcal/mol with respect to the parent compound ZZW-115. The same variations in the affinity were obtained by cross-docking ZZW-142 into the equilibrated binding pockets containing ZZW-115, or *vice versa*. As shown in Figure 2, differences were due to the arrangement in the conformation of the alkylated chains of the two compounds, while the position of the scaffold remained essentially unchanged. Results were not equally favorable when methylene groups were added in a distal position with respect to the amine group of ZZW-115 (as in compound ZZW-129), because a slightly worse binding energy was obtained (+0.1 kcal/mol).

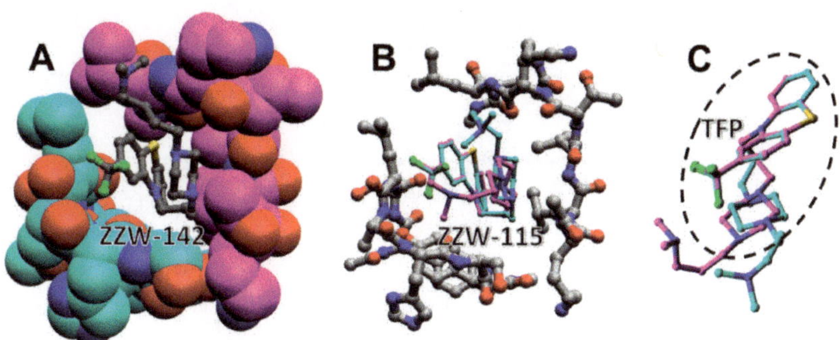

Figure 2. Docking of compounds into the simulated binding pocket of NUPR1: (**A**) Compound ZZW-142 bound into an example of transient pocket formed by the 7-residue regions of NUPR1 centered around Ala33 (cyan) and Thr68 (magenta). (**B**) Cross-docking of compound ZZW-115 into the same binding pocket, showing two different conformations (cyan and magenta). (**C**) Details of the two conformations of ZZW-115 from a different orientation, highlighting the TFP scaffold and the differences in the alkylated tail conformation.

It is worth to note that the small energetic differences obtained are close to the uncertainty in the evaluation provided by the empiric scoring function, which is on the order of 0.1–0.2 kcal/mol [43]. More importantly, as regards the protein pockets used to model the binding: (i) they have a larger conformational variety (or "fuzziness", in the terminology commonly used for IDPs) compared to well-defined binding sites, such as those that could be obtained from experimental techniques for well-folded proteins (e.g., X-ray crystallography or NMR) or by homology modelling; and (ii) they cannot be claimed to span an accurate statistical ensemble, whose determination is beyond the current possibilities of both experimental and theoretical methods. These factors restrain the predictive power of the computational techniques and, overall, demonstrate the limitation of LBDD in helping to design IDP inhibitors. In particular, no reliable predictions could be obtained when bulkier chemical moieties were added (as in compounds ZZW-130, ZZW-131 and ZZW-132), because uncertainties in the determination of the binding score were too large (on the order of 1 kcal/mol), and the alkylated tail became too dissimilar to allow a direct comparison with ZZW-115 by using cross-docking.

3.2. Synthesis of the ZZW-115-Derived Compounds

We have designed nine 10-(3-(4-ethylpiperazin-1-yl) propyl-10H-phenotiazine derivatives, all modified with different groups on the ethylpiperazine chain. Similar to the synthesis of ZZW-115 [26], the reaction between phenothiazine compound **a** and various piperazine derivatives could produce the ZZW-115-derived compounds in a yield that ranged from 60% to 80% (Scheme S1). The chemical structures and the purities of these compounds were characterized by ^{1}H-NMR, ^{13}C-NMR, MS and HPLC.

3.3. Isothermal Titration Calorimetry in the Presence of ZZW-115-Derived Compounds

The interaction between ZZW-115-derived compounds and NUPR1 was studied by using ITC, the gold-standard technique in binding-affinity determination, to test our theoretical prediction. The affinity, enthalpy, and stoichiometry of binding were determined at 25 °C. From the thermogram (thermal power as a function of time), the binding isotherm (ligand-normalized heat as a function of the molar ratio) was obtained through integration of the individual heat effect associated with injection of each ligand solution (Figure 3). We used a nonlinear least-squares regression analysis employing a model that considered a single binding site in NUPR1 to estimate binding parameters such as the dissociation constant and the binding enthalpy.

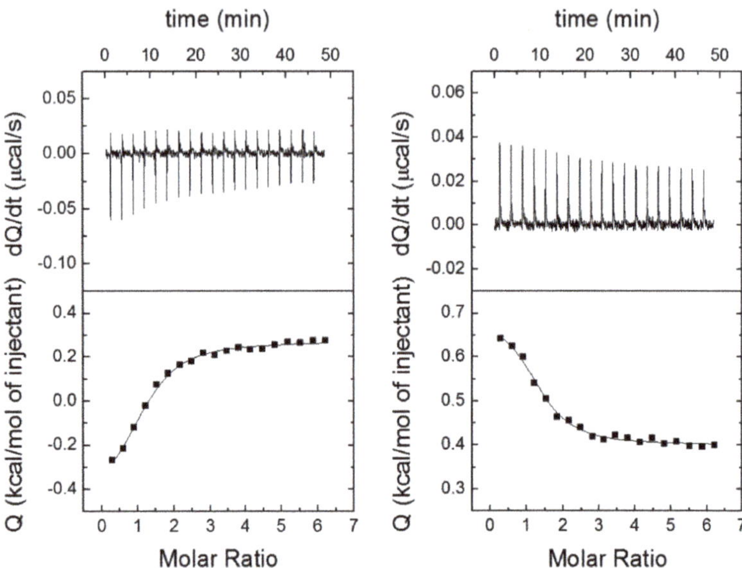

Figure 3. Calorimetric titrations corresponding to the interaction of ZZW-129 (left) and ZZW-142 (right) with NUPR1: Thermograms (thermal power as a function of time) are shown on the top, and binding isotherms (ligand-normalized heat effects as a function of molar ratio) are shown on the bottom. Experiments were performed at 25 °C in sodium phosphate 20 mM, pH 7, 2% DMSO, with 20 µM NUPR1 in the calorimetric cell and 200 µM compound in the titrating syringe, using an Auto-iTC200 instrument (MicroCal-Malvern Panalytical).

All compounds exhibited dissociation constants in the low micromolar range (Table 1). The interaction is dominated by the entropic contribution, whereas the enthalpic contribution is negligible; all the ΔH values are within a very narrow range of 1 kcal/mol, and with either a negative or positive sign that may depending on small differences upon the mode of interaction. In such a situation, the background heat effect for the injection is large, because even a small difference in DMSO concentration largely modulates this quantity. Nevertheless, this effect does not dominate the thermogram and the interaction parameters could be accurately estimated from the fits of the ITC curves (see again Figure 3). The affinity data do not show in general a clear trend, although it can be noted that the most favorable values have the largest entropic contributions.

Table 1. Thermodynamic parameters of the binding reaction between NUPR1 and the compounds. [a] Obtained from ITC measurements; errors are estimated from the fitting and standard thermodynamic relationships ($K_d = 1/K_a$; $\Delta G = -RT \ln K_a$; $\Delta G = \Delta H - T\Delta S$). [b] Data taken from [26].

Compound	K_a (M^{-1}) $\times 10^5$	K_d (μM) (=1/K_a) [a]	ΔG (kcal mol^{-1}) [a]	ΔH (kcal mol^{-1}) [a]	$-T\Delta S$ (kcal mol^{-1}) [a]	N [a]
ZZW-129	3.1 ± 0.3	3.2 ± 0.3	-7.5 ± 0.1	-0.7 ± 0.4	-6.8 ± 0.4	1.2 ± 0.1
ZZW-130	6.4 ± 0.4	1.6 ± 0.1	-7.9 ± 0.1	0.3 ± 0.3	-8.2 ± 0.3	1.4 ± 0.1
ZZW-131	4.7 ± 0.3	2.1 ± 0.1	-7.7 ± 0.1	0.5 ± 0.3	-8.2 ± 0.3	0.8 ± 0.1
ZZW-132	5.8 ± 0.4	1.7 ± 0.1	-7.9 ± 0.1	0.3 ± 0.3	-8.2 ± 0.3	1.2 ± 0.1
ZZW-142	4.9 ± 0.3	2.0 ± 0.2	-7.8 ± 0.1	0.3 ± 0.3	-8.1 ± 0.3	1.3 ± 0.1
ZZW-143	0.5 ± 0.1	20 ± 4	-6.4 ± 0.1	0.8 ± 0.4	-7.2 ± 0.4	1.2 ± 0.1
ZZW-144	0.85 ± 0.8	12 ± 1	-6.7 ± 0.1	0.5 ± 0.5	-7.2 ± 0.5	1.2 ± 0.1
ZZW-145	0.82 ± 0.9	11 ± 1	-6.7 ± 0.1	0.7 ± 0.5	-7.4 ± 0.5	1.2 ± 0.1
ZZW-148	1.0 ± 0.1	9.6 ± 1.0	-6.8 ± 0.1	0.7 ± 0.5	-7.5 ± 0.5	1.2 ± 0.1
ZZW-115 [b]	4.7 ± 0.4	2.1 ± 0.2	-7.7 ± 0.1	-0.4 ± 0.3	-7.3 ± 0.3	0.9 ± 0.1
TFP [b]	1.9 ± 0.2	5.2 ± 0.6	-7.2 ± 0.1	-1.1 ± 0.4	-6.1 ± 0.4	1.0 ± 0.1

Among the compounds tested, ZZW-130 and ZZW-132 were the strongest binders, with a dissociation constant close to 1.7 µM (Figure 3), slightly smaller than that of ZZW-115. These results suggest that the addition of a heterocyclic ring at the terminus of the ethylene moiety hanging from the piperazine ring does not affect the binding affinity of the compounds to NUPR1. That is, the protein is flexible enough to accommodate a bulky ring around its hot spot regions.

In contrast, the affinity constants changed dramatically when the chain hanging from the piperazine ring was enlarged from an ethyl to a propyl group (i.e., a methylene moiety was added) and, in addition, a large ring was added at the terminal end of the new propyl. For instance, ZZW-142 had slightly more favorable affinity for NUPR1 ($K_d = 2.0$ µM) than the corresponding parental compound, ZZW-115 (Table 1), suggesting that the addition of a methylene only slightly affected the binding to NUPR1. However, when a heterocyclic ring was incorporated to the terminus of the propyl group, the affinity decreased (Table 1). It is likely that the addition of the propyl group, together with the incorporation of the heterocyclic ring, cannot be accommodated by NUPR1 due to the larger size of the resulting compound. Thus, compounds ZZW-143, ZZW-144, ZZW-145, and ZZW-148 displayed the weakest affinity for NUPR1. The very similar interaction enthalpy observed for most of the compounds suggest that the loss in binding affinity for those containing a longer propyl chain may be related to a conformational entropy loss from the compound stemming from conformationally constraining an additional rotatable bond. This observation is in agreement with the results of our molecular simulations, which showed that the scaffold of TFP essentially dictates the binding affinity of the various ZZW-115-derived compounds, and contributes to explain the difficulties in predicting in silico the fine details of the binding of these ligands, because entropic quantities are notoriously more complex to obtain compared to enthalpic ones by using computational techniques.

3.4. Effects of ZZW-115-Derived Compounds on Nuclear Translocation of NUPR1

In our previous work, we have shown that treatment with ZZW-115 inhibited almost completely the translocation of NUPR1 from the cytoplasm to the nucleus by competing with importins in binding their nuclear localization sequence [27]. Here, we observed that all the ZZW-115-derived compounds were capable of hampering NUPR1 nuclear translocation with a higher efficiency than TFP (Figure 4). ZZW-143 and ZZW-145 had an ability to arrest NUPR1 nuclear translocation similar to that of ZZW-115, followed by the efficiency of ZZW-144. This result is somewhat surprising, as compounds ZZW-143 and ZZW-145 had a weaker affinity in vitro for NUPR1 than the rest of the compounds (Table 1). The other ZZW-115-derived compounds had a lower capacity to hamper NUPR1 translocation (Figure 4 and Figure S1). In particular, the worst effect in hampering the nuclear translocation efficiency corresponded to compounds ZZW-132 and ZZW-142.

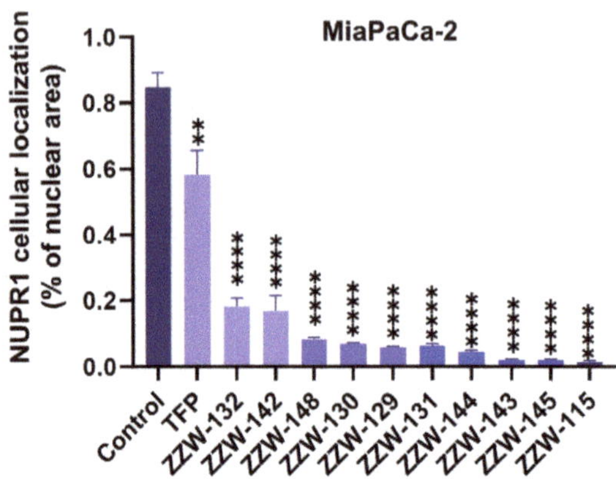

Figure 4. ZZW-115-derived compounds inhibited NUPR1 nuclear translocation. MiaPaCa-2 cells were treated with representative ZZW-115-derived compounds (3 μM) for 6 h. Immunofluorescence with rabbit anti-NUPR1 primary antibody and Alexa 488–labeled goat anti-rabbit secondary antibody were used to reveal the localization of the protein (n = 3). DAPI staining was used to detect cell nuclei, and it was combined with the Alexa 488 fluorescence in the merged panel. The *y*-axis indicates the fraction of area within the nucleus where the presence of NUPR1 was observed. ** $p < 0.01$; **** $p < 0.0001$ (1-way ANOVA, Tukey's post hoc test).

3.5. Treatment with ZZW-115-Derived Compounds Sensitizes Cancer Cells to Genotoxic-induced DNA Damage

As NUPR1 is involved in DNA-damage stimulus processes [13,27], we had shown in our previous work that treatment with ZZW-115 sensitizes cancer cells to genotoxic-induced DNA damage [27]. Hence, we hypothesized that ZZW-115-derived compounds could have a similar effect. However, to our surprise, we observed that the ZZW-115-derived compounds showed lower improvement in sensitizing cancer cells to DNA damage induced by 5-FU than the original compound ZZW-115, as indicated by the decrease in the number of foci per cell observed in our experiments (Figures 5 and S2). The sole compound showing a sensitizing effect similar to ZZW-115 was ZZW-145 (which had one of the lowest affinities for NUPR1 in vitro, Table 1). On the other hand, the worst improvement in sensitizing cancer cells to 5-FU occurred for compounds ZZW-132 and ZZW-142 (Figure 5), which showed results similar to the effect caused by the sole action of 5-FU. Therefore, the smallest sensitizing effect was observed for the same two compounds hampering nuclear translocation of NUPR1.

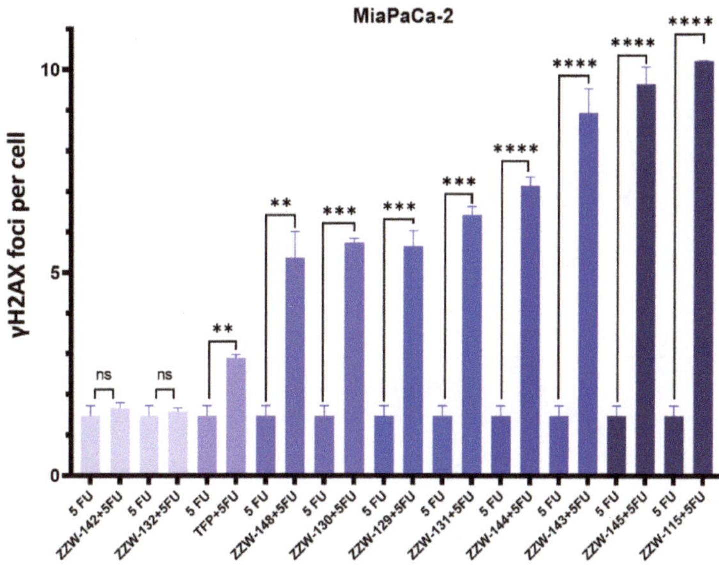

Figure 5. NUPR1 inhibition by ZZW-115-derived compounds potentiated the efficacy of 5-FU in MiaPaCa-2 cells. The efficacy of 5-FU to generate DNA breaks in primary PDAC cells and the boosting effect of ZZW-115 was evaluated by γH2AX immunofluorescence staining. Quantifications of 3 independent experiments were used to evaluate the statistical significance, and they are shown as graphics. ** $p < 0.01$; *** $p < 0.001$; **** $p < 0.0001$ (1-way ANOVA, Tukey's post hoc test). Data represent mean $\pm$ SEM, n = 3.

3.6. Effects of ZZW-115-Derived Compounds on Pancreatic Cancer Cell Growth

Treatment of MiaPaCa-2 (a traditional cell line), PDAC001T, PDAC021T, PDAC087T and PDAC115T (basal subtype), PDAC089T (derived from a liver metastasis), and PDAC012T, PDAC081T, PDAC082T, PDAC088T (classical subtype) cells with the ZZW-115-derived compounds showed different effects on the growth of pancreatic cancer cells (Table 2). The results were compared with those previously obtained by TFP and ZZW-115 [26], used as controls. ZZW-115 and most of its derived compounds were found to be 3 to 10 times more efficient in killing cancer cells than the treatment with TFP (and other TFP-derived compounds [26]) in MiaPaCa-2 cells (Figure 6, Table 2). However, most of the compounds showed values of IC$_{50}$ similar to that of ZZW-115 in MiaPaCa-2 cells. Only ZZW-132 and ZZW-142 showed worse activity (having IC$_{50}$ values in the range of 1 μM), although the affinities of both observed in vitro were similar to that of ZZW-115 (Table 1). Taken together, the overall results obtained with MiaPaCa-2 cells demonstrate that the chemical modifications introduced did not seem to increase appreciably the anticancer activity when compared to that of ZZW-115, except for two compounds. More importantly, we did not observe any direct relationship between the affinity measured in vitro by ITC and the anticancer activity of the same compound, neither, in general terms, among the effectivity of the drugs in the different biological activities measured.

Table 2. IC$_{50}$ (in µM) for the different assayed cell-lines with the ZZW-115-derived compounds.

	PDAC001T	PDAC012T	PDAC021T	PDAC081T	PDAC082T	PDAC087T	PDAC088T	PDAC089T	PDAC115T	MiaPaCa-2
ZZW-129	1.94	2.77	3.41	0.86	3.56	4.28	7.28	7.90	2.46	0.33
ZZW-130	2.45	4.18	3.81	1.12	4.33	4.95	6.75	6.03	2.91	0.35
ZZW-131	3.98	5.25	6.47	0.97	6.49	4.43	8.11	10.59	5.75	0.30
ZZW-132	12.47	12.21	16.50	1.91	21.10	16.55	23.63	29.10	16.06	1.04
ZZW-142	11.93	12.78	17.01	2.27	22.97	16.55	23.92	22.46	14.70	0.95
ZZW-143	2.31	2.18	2.57	0.97	3.55	2.22	5.94	6.12	2.50	0.29
ZZW-144	2.51	2.51	3.69	0.98	3.04	3.57	5.80	8.38	3.22	0.26
ZZW-145	2.36	2.36	2.33	0.78	2.33	2.37	5.00	6.22	2.44	0.19
ZZW-148	3.75	3.70	4.35	1.11	4.10	6.57	10.73	14.77	4.37	0.47
ZZW-115	1.84	2.44	2.54	0.84	2.36	3.72	5.29	4.72	2.21	0.32
TFP	12.48	20.39	23.51	2.48	21.47	7.80	15.90	8.09	10.12	9.28

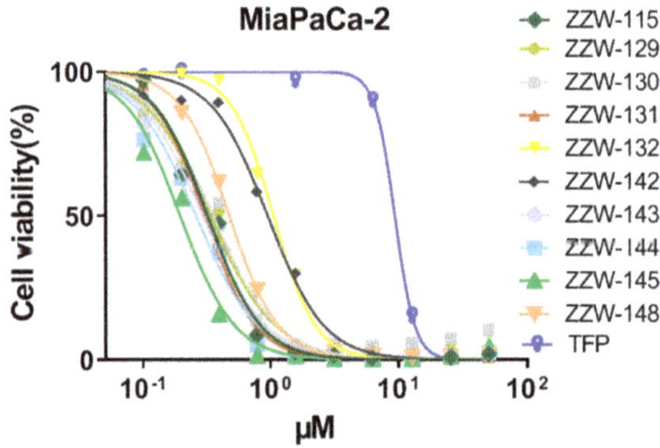

Figure 6. ZZW-115-derived compounds have antitumoral effect: Viability of MiaPaCa-2 cells upon a treatment for 72 h with the ZZW-115-derived compounds.

For the rest of the assayed cells (Table 2), in broad terms, most of the IC$_{50}$ values were similar to those of ZZW-115, and lower than those of TFP (i.e., they had a better anti-cancer activity). Only ZZW-132 (as it happened with MiaPaca-2 cells) and ZZW-142 were found to show worse IC$_{50}$ values. These results are especially surprising for ZZW-132, as its affinity measured in vitro by ITC was among the most favorable (Table 1).

4. Discussion

In this work, we sought to use an indirect design to obtain new compounds targeting the totally unstructured tumorigenic protein NUPR1, building on former successful efforts that had led to the identification of the anti-cancer drug ZZW-115 [24]. Our attempts to further modify this compound were based on previous results obtained starting from the chemical scaffold of TFP, which has the ability to bind NUPR1 [25], and on the chemical features and biological effects of ZZW-115 and other TFP-derived analogs [26]. In our previous studies, we had also used NMR to further characterize the binding of ZZW-115, TFP, and other compounds; however, such technique always identified small variations in the intensity of peaks of residues around the above mentioned hot spots [25,26], which were in complete agreement with the corresponding MD simulations. Such previous accord between MD and NMR studies has prompted us to use only the former technique in the present study on the derivatives of ZZW-115.

We could not develop inhibitors more effective than ZZW-115 due the difficulties in obtaining reliable results by using solely a rational LBDD approach. In fact, the differences obtained by adding in silico small chemical moieties, such as single methylene group in the alkyl chain attached to the piperazine ring of TFP, resulted in very small modifications of the binding affinity, as obtained from the simulation experiments. Moreover, for larger chemical moieties the comparison with ZZW-115 became unclear, and no reliable result could be obtained. Furthermore, at this stage we cannot unambiguously rule out the possibility of binding of any of the compounds to other proteins or other biomolecules. However, studies carried out with previous compounds (TFP, ZZW-115 and its first series of derivatives) on NUPR1-deficient mice did not result in the inhibition of cancer growth [25,26], suggesting that the main target for these drugs is in fact NUPR1.

The modifications introduced to obtain ZZW-115 derivatives were tested by ITC measurements in vitro (Table 1). It was observed that a larger affinity (i.e., smaller dissociation constant) generally corresponded to compounds in which a smaller moiety with an ethylene chain was present in the piperazine ring of the compound. An exception to this behavior was observed for compound ZZW-142, which was the sole compound for which the simulation results gave a more reliable prediction. The computational results and the ITC measurements both indicate that there was no additive effect on the binding affinity obtained by introducing two different chemical groups to the same scaffold. In fact, the introduction of a methylene group slightly improved the binding, in the presence of the amine group in the alkyl tail (as visible in the comparison between the parent compound ZZW-115 and the derived analog ZZW-142); whereas the same modification strongly decreased the binding affinity of compound ZZW-130, which is the one with the best binding affinity for NUPR1 found in the experiment (leading to compound ZZW-144, which in contrast has a poor affinity). Although the non-additivity of molecular interactions is a well-known phenomenon that is recognized to hamper even the structure-based drug design of ligands targeting well-structured binding sites [44], this is likely to pose even greater difficulties in the case of ligand-based approaches against IDPs.

Another finding in our drug design effort was that our biological results (Figures 4–6) indicated that a larger affinity, as measured by ITC, does not necessarily translate into a more favorable inhibitory activity. In fact, the results obtained in cellulo clearly indicated no clear correlation between ITC data and specific improvement in either of the following aspects: (i) better hampering of nuclear translocation of NUPR1; (ii) superior sensitizing of cells against DNA damage mediated by NUPR1; and (iii) more evident anticancer cell activity. In particular, the results observed as regards these three biological indicators were the poorest for ZZW-132 and ZZW-142, although such compounds are among the ones with the most favorable binding affinity for NUPR1 observed from ITC measurements (Table 1). This finding supports the notion that identifying the relationship between the chemical structure of compounds and their resulting biological activity, which is very difficult in the case of well-folded structures, is even more arduous for IDPs. At the moment, it is not clear how to further modify ZZW-115 to obtain better biological effects, although a number of examples reported in the literature of drug design against IDPs are encouraging in this sense [45,46].

Although it is difficult to identify the origin of the lack of correlation between our biophysical and biological results, it is interesting to suggest at least some possible explanations. First, NUPR1 is an IDP with a totally unfolded structure and it remains completely disordered to carry out its different functions, even in the presence of its natural partners (such as DNA or other proteins [12,15]) or when bound to TFP or TFP-derived drugs [25,26]. A compound binding with a high affinity may freeze the inherent internal flexibility of NUPR1, impacting its activity in other ways. It is worth to note that a residual flexibility has been shown to be necessary to carry out the function of a number of proteins: for instance, neurotensin needs to maintain its flexibility around the crucial residue Tyr11 when the protein is bound to neurotensin receptor 1 [47], and the same applies for a whole nascent helix of dynein when it is bound to dynactin [48]. As a second possible explanation for the biological effects we observed, it has been recently argued that an effective strategy for drug discovery in IDPs relies in exploiting the entropic contributions due to the protein chain flexibility, which would increase the affinity towards ligands by shifting the population of disordered conformations [49,50]. This approach (to increase disorder-upon-binding by shifting the fraction of disordered conformations of an IDP) has been successfully used in some cases, such as in the design of a drug targeting the kinase activation loop of Bcr-Abl [51], and can even work for well-folded proteins, as it helped in the identification of carbohydrate ligands targeting galectin-3 [52]; furthermore, such approach was instrumental for the entropy gain related to the lower susceptibility to resistance-associated mutations of certain inhibitors of the HIV-1 protease [53]. In the case of NUPR1, compounds ZZW-132 and ZZW-142 may shift the conformational populations of NUPR1 by increasing its flexibility, but such a shift could reduce the fraction of conformers that are active to perform some tasks (for instance, to allow NUPR1 translocation through the nuclear pore). Whether the explanation is either of the two proposed here or a different one, such effects are likely to have a role in the drug design against IDPs in general, and not only for NUPR1. These concerns supplement those that already exist in any drug development attempt, including those against well-folded protein targets, such as the fact that differences might also be due to other causes that include: (i) variations in the cellular uptake of the modified compounds; (ii) alterations in their cellular metabolism; (iii) the relative favorable binding affinity of the designed compounds for other proteins or biological substrates [54]; or even (iv) the formation of aggregates (oligomeric species of NUPR1, in this case) triggered by the presence of the bound compounds. Thus, the biological effect of a given compound (primary effects plus side-effects) will depend on the binding affinity towards the primary target, the effective concentration of the compound at the appropriate target location, and the binding affinities towards unwanted targets.

5. Conclusions

In this work we have pursued the design of novel inhibitors of the tumorigenic protein NUPR1, and suggested that a subtle compromise between improving ligand affinity and maintaining protein functionality may be required. Whereas increasing the affinity towards NUPR1 is important to obtain a more potent drug, the functional pathways this protein belongs to should not be exceedingly perturbed, in order to maintain their effectiveness in performing some of their biological roles. This poses significant limits in the possibility of relying solely on the most traditional theoretical modelling techniques, including molecular docking and MD. A counterintuitive consequence is that, for disordered proteins performing multiple functions within cells, an improved inhibitory effect may be attained by investigating in cellulo the properties of compounds with a comparable—or even slightly reduced—binding affinity with respect to a lead drug compound (such as ZZW-115, in our case). Therefore, we provide a direct evidence that also for IDPs the best drug is not necessarily the best binder. This finding enlarges the possibility in the number of active molecules that can be hoped to be effective against IDPs, encouraging the exploration of regions of the chemical space of drug compounds that may be currently neglected or underrated.

Supplementary Materials: The following are available online at https://www.mdpi.com/article/10.3390/biom11101453/s1. Scheme S1: Synthesis of ZZW-115-derived compounds. Figure S1: capacity of ZZW-115-derived compounds to perform nuclear translocation. Figure S2: capacity of ZZW-115-derived compounds to sensitize 5-FU induced DNA damage. The ^{1}H, ^{13}C-NMR spectra and HPLC spectra of the compounds are also included.

Author Contributions: Conceptualization and design: B.R., P.S.-C., A.V.-C., O.A., L.P., J.L.N., Y.X. and J.L.I.; investigation and formal analysis: B.R., W.L., P.S.-C., A.V.-C., O.A. and J.L.I.; synthesis of the compounds and analysis of spectra and chromatograms: Z.Z. and Y.X. All authors have read and agreed to the published version of the manuscript.

Funding: This work was supported by Spanish Ministry of Economy and Competitiveness and European ERDF Funds (MCIU/AEI/FEDER, EU) [RTI2018-097991-B-I00 to J.L.N.]; La Ligue Contre le Cancer, INCa, Canceropole PACA and INSERM to JLI; Fondo de Investigaciones Sanitarias from Instituto de Salud Carlos III, and European Union (ERDF/ESF, 'Investing in your future') (PI18/00349 to O.A.); Diputación General de Aragón ('Protein Targets and Bioactive Compounds Group' E45_20R to A.V.-C., 'Digestive Pathology Group' B25_20R to O.A.); Centro de Investigación Biomédica en Red en Enfermedades Hepáticas y Digestivas (CIBERehd) (O.A. and A.V.-C.); CAI YUANPEI Scholarship (201906050187) to Y.X.; and Jeunes Talents France-Chine Program (JTFC) to Y.X. P.S.-C. was supported by Fondation de France; and W.L. by China Scholarship Council.

Data Availability Statement: Materials are available from any of the corresponding authors upon reasonable request.

Acknowledgments: B.R. acknowledges the kind use of computational resources by the European Magnetic Resonance Center (CERM), Sesto Fiorentino (Florence), Italy. PyMol version 2.1 (available at http://www.pymol.org/pymol, accessed on 2 October 2021) from Schrödinger LLC was used for molecular graphics.

Conflicts of Interest: Some of the authors of this work (B.R., P.S.-C., A.V.-C., O.A., L.P., J.L.N., Y.X. and J.L.I.) hold the co-authorship of a patent on the use of ZZW-115 and other TFP-derived compounds as inhibitors of NUPR1 with anticancer effects.

References

1. Babu, M.M.; van der Lee, R.; de Groot, N.S.; Gsponer, J. Intrinsically Disordered Proteins: Regulation and Disease. *Curr. Opin. Struct. Biol.* **2011**, *21*, 432–440. [CrossRef] [PubMed]
2. Xie, H.; Vucetic, S.; Iakoucheva, L.M.; Oldfield, C.J.; Dunker, A.K.; Uversky, V.N.; Obradovic, Z. Functional Anthology of Intrinsic Disorder. 1. Biological Processes and Functions of Proteins with Long Disordered Regions. *J. Proteome Res.* **2007**, *6*, 1882–1898. [CrossRef] [PubMed]
3. Berlow, R.B.; Dyson, H.J.; Wright, P.E. Expanding the Paradigm: Intrinsically Disordered Proteins and Allosteric Regulation. *J. Mol. Biol.* **2018**, *430*, 2309–2320. [CrossRef] [PubMed]
4. Gsponer, J.; Futschik, M.E.; Teichmann, S.A.; Babu, M.M. Tight Regulation of Unstructured Proteins: From Transcript Synthesis to Protein Degradation. *Science* **2008**, *322*, 1365–1368. [CrossRef] [PubMed]
5. Banani, S.F.; Lee, H.O.; Hyman, A.A.; Rosen, M.K. Biomolecular Condensates: Organizers of Cellular Biochemistry. *Nat. Rev. Mol. Cell Biol.* **2017**, *18*, 285–298. [CrossRef] [PubMed]
6. Hu, G.; Wu, Z.; Uversky, V.N.; Kurgan, L. Functional Analysis of Human Hub Proteins and Their Interactors Involved in the Intrinsic Disorder-Enriched Interactions. *Int. J. Mol. Sci.* **2017**, *18*, 2761. [CrossRef]
7. Mallo, G.V.; Fiedler, F.; Calvo, E.L.; Ortiz, E.M.; Vasseur, S.; Keim, V.; Morisset, J.; Iovanna, J.L. Cloning and Expression of the Rat P8 CDNA, a New Gene Activated in Pancreas during the Acute Phase of Pancreatitis, Pancreatic Development, and Regeneration, and Which Promotes Cellular Growth. *J. Biol. Chem.* **1997**, *272*, 32360–32369. [CrossRef]
8. Cano, C.E.; Hamidi, T.; Sandi, M.J.; Iovanna, J.L. Nupr1: The Swiss-Knife of Cancer. *J. Cell Physiol.* **2011**, *226*, 1439–1443. [CrossRef]
9. Goruppi, S.; Iovanna, J.L. Stress-Inducible Protein P8 Is Involved in Several Physiological and Pathological Processes. *J. Biol. Chem.* **2010**, *285*, 1577–1581. [CrossRef] [PubMed]
10. Chowdhury, U.R.; Samant, R.S.; Fodstad, O.; Shevde, L.A. Emerging Role of Nuclear Protein 1 (NUPR1) in Cancer Biology. *Cancer Metastasis Rev.* **2009**, *28*, 225–232. [CrossRef]
11. Santofimia-Castaño, P.; Xia, Y.; Peng, L.; Velázquez-Campoy, A.; Abián, O.; Lan, W.; Lomberk, G.; Urrutia, R.; Rizzuti, B.; Soubeyran, P.; et al. Targeting the Stress-Induced Protein NUPR1 to Treat Pancreatic Adenocarcinoma. *Cells* **2019**, *8*, 1453. [CrossRef] [PubMed]
12. Malicet, C.; Giroux, V.; Vasseur, S.; Dagorn, J.C.; Neira, J.L.; Iovanna, J.L. Regulation of Apoptosis by the P8/Prothymosin Alpha Complex. *Proc Natl Acad Sci USA* **2006**, *103*, 2671–2676. [CrossRef]
13. Aguado-Llera, D.; Hamidi, T.; Doménech, R.; Pantoja-Uceda, D.; Gironella, M.; Santoro, J.; Velázquez-Campoy, A.; Neira, J.L.; Iovanna, J.L. Deciphering the Binding between Nupr1 and MSL1 and Their DNA-Repairing Activity. *PLoS ONE* **2013**, *8*, e78101. [CrossRef] [PubMed]
14. Encinar, J.A.; Mallo, G.V.; Mizyrycki, C.; Giono, L.; Gonzalez-Ros, J.M.; Rico, M.; Cánepa, E.; Moreno, S.; Neira, J.L.; Iovanna, J.L. Human P8 Is a HMG-I/Y-like Protein with DNA Binding Activity Enhanced by Phosphorylation. *J. Biol. Chem.* **2001**, *276*, 2742–2751. [CrossRef] [PubMed]
15. Santofimia-Castaño, P.; Rizzuti, B.; Pey, Á.L.; Soubeyran, P.; Vidal, M.; Urrutia, R.; Iovanna, J.L.; Neira, J.L. Intrinsically Disordered Chromatin Protein NUPR1 Binds to the C-Terminal Region of Polycomb RING1B. *Proc. Natl. Acad. Sci. USA* **2017**, *114*, E6332–E6341. [CrossRef] [PubMed]
16. Sandi, M.J.; Hamidi, T.; Malicet, C.; Cano, C.; Loncle, C.; Pierres, A.; Dagorn, J.C.; Iovanna, J.L. P8 Expression Controls Pancreatic Cancer Cell Migration, Invasion, Adhesion, and Tumorigenesis. *J. Cell Physiol.* **2011**, *226*, 3442–3451. [CrossRef]
17. Vasseur, S.; Hoffmeister, A.; Garcia, S.; Bagnis, C.; Dagorn, J.-C.; Iovanna, J.L. P8 Is Critical for Tumour Development Induced by RasV12 Mutated Protein and E1A Oncogene. *EMBO Rep.* **2002**, *3*, 165–170. [CrossRef]
18. Emma, M.R.; Iovanna, J.L.; Bachvarov, D.; Puleio, R.; Loria, G.R.; Augello, G.; Candido, S.; Libra, M.; Gulino, A.; Cancila, V.; et al. NUPR1, a New Target in Liver Cancer: Implication in Controlling Cell Growth, Migration, Invasion and Sorafenib Resistance. *Cell Death Dis.* **2016**, *7*, e2269. [CrossRef]

19. Kim, K.-S.; Jin, D.-I.; Yoon, S.; Baek, S.-Y.; Kim, B.-S.; Oh, S.-O. Expression and Roles of NUPR1 in Cholangiocarcinoma Cells. *Anat. Cell Biol.* **2012**, *45*, 17–25. [CrossRef]

20. Guo, X.; Wang, W.; Hu, J.; Feng, K.; Pan, Y.; Zhang, L.; Feng, Y. Lentivirus-Mediated RNAi Knockdown of NUPR1 Inhibits Human Nonsmall Cell Lung Cancer Growth in Vitro and in Vivo. *Anat. Rec. (Hoboken)* **2012**, *295*, 2114–2121. [CrossRef]

21. Li, J.; Ren, S.; Liu, Y.; Lian, Z.; Dong, B.; Yao, Y.; Xu, Y. Knockdown of NUPR1 Inhibits the Proliferation of Glioblastoma Cells via ERK1/2, P38 MAPK and Caspase-3. *J. Neurooncol.* **2017**, *132*, 15–26. [CrossRef]

22. Zhou, C.; Xu, J.; Lin, J.; Lin, R.; Chen, K.; Kong, J.; Shui, X. Long Noncoding RNA FEZF1-AS1 Promotes Osteosarcoma Progression by Regulating the MiR-4443/NUPR1 Axis. *Oncol. Res.* **2018**, *26*, 1335–1343. [CrossRef]

23. Zeng, C.; Li, X.; Li, A.; Yi, B.; Peng, X.; Huang, X.; Chen, J. Knockdown of NUPR1 Inhibits the Growth of U266 and RPMI8226 Multiple Myeloma Cell Lines via Activating PTEN and Caspase Activation–dependent Apoptosis. *Oncol. Rep.* **2018**, *40*, 1487–1494. [CrossRef]

24. Santofimia-Castaño, P.; Rizzuti, B.; Xia, Y.; Abian, O.; Peng, L.; Velázquez-Campoy, A.; Iovanna, J.L.; Neira, J.L. Designing and Repurposing Drugs to Target Intrinsically Disordered Proteins for Cancer Treatment: Using NUPR1 as a Paradigm. *Mol. Cell Oncol.* **2019**, *6*, e1612678. [CrossRef] [PubMed]

25. Neira, J.L.; Bintz, J.; Arruebo, M.; Rizzuti, B.; Bonacci, T.; Vega, S.; Lanas, A.; Velázquez-Campoy, A.; Iovanna, J.L.; Abián, O. Identification of a Drug Targeting an Intrinsically Disordered Protein Involved in Pancreatic Adenocarcinoma. *Sci. Rep.* **2017**, *7*, 39732. [CrossRef] [PubMed]

26. Santofimia-Castaño, P.; Xia, Y.; Lan, W.; Zhou, Z.; Huang, C.; Peng, L.; Soubeyran, P.; Velázquez-Campoy, A.; Abián, O.; Rizzuti, B.; et al. Ligand-Based Design Identifies a Potent NUPR1 Inhibitor Exerting Anticancer Activity via Necroptosis. *J. Clin. Investig.* **2019**, *129*, 2500–2513. [CrossRef] [PubMed]

27. Lan, W.; Santofimia-Castaño, P.; Swayden, M.; Xia, Y.; Zhou, Z.; Audebert, S.; Camoin, L.; Huang, C.; Peng, L.; Jiménez-Alesanco, A.; et al. ZZW-115-Dependent Inhibition of NUPR1 Nuclear Translocation Sensitizes Cancer Cells to Genotoxic Agents. *JCI Insight* **2020**, *5*, e138117. [CrossRef]

28. Leelananda, S.P.; Lindert, S. Computational Methods in Drug Discovery. *Beilstein J. Org. Chem.* **2016**, *12*, 2694–2718. [CrossRef]

29. Macalino, S.J.Y.; Gosu, V.; Hong, S.; Choi, S. Role of Computer-Aided Drug Design in Modern Drug Discovery. *Arch. Pharm. Res.* **2015**, *38*, 1686–1701. [CrossRef]

30. Rizzuti, B.; Grande, F. Virtual screening in drug discovery: A precious tool for a still-demanding challenge. In *Protein Homeostasis Diseases*; Elsevier BV: Amsterdam, The Netherlands, 2020; pp. 309–327.

31. Gill, S.C.; von Hippel, P.H. Calculation of Protein Extinction Coefficients from Amino Acid Sequence Data. *Anal. Biochem.* **1989**, *182*, 319–326. [CrossRef]

32. Trott, O.; Olson, A.J. AutoDock Vina: Improving the Speed and Accuracy of Docking with a New Scoring Function, Efficient Optimization, and Multithreading. *J. Comput. Chem.* **2010**, *31*, 455–461. [CrossRef]

33. Hess, B.; Kutzner, C.; van der Spoel, D.; Lindahl, E. GROMACS 4: Algorithms for Highly Efficient, Load-Balanced, and Scalable Molecular Simulation. *J. Chem. Theory Comput.* **2008**, *4*, 435–447. [CrossRef]

34. Lindorff-Larsen, K.; Piana, S.; Palmo, K.; Maragakis, P.; Klepeis, J.L.; Dror, R.O.; Shaw, D.E. Improved Side-Chain Torsion Potentials for the Amber Ff99SB Protein Force Field. *Proteins* **2010**, *78*, 1950–1958. [CrossRef] [PubMed]

35. Wang, J.; Wolf, R.M.; Caldwell, J.W.; Kollman, P.A.; Case, D.A. Development and Testing of a General Amber Force Field. *J. Comput. Chem.* **2004**, *25*, 1157–1174. [CrossRef]

36. Piana, S.; Donchev, A.G.; Robustelli, P.; Shaw, D.E. Water Dispersion Interactions Strongly Influence Simulated Structural Properties of Disordered Protein States. *J. Phys. Chem. B* **2015**, *119*, 5113–5123. [CrossRef] [PubMed]

37. Guglielmelli, A.; Rizzuti, B.; Guzzi, R. Stereoselective and Domain-Specific Effects of Ibuprofen on the Thermal Stability of Human Serum Albumin. *Eur. J. Pharm. Sci.* **2018**, *112*, 122–131. [CrossRef] [PubMed]

38. Neira, J.L.; Rizzuti, B.; Iovanna, J.L. Determinants of the pK_a Values of Ionizable Residues in an Intrinsically Disordered Protein. *Arch. Biochem. Biophys.* **2016**, *598*, 18–27. [CrossRef]

39. Nicolle, R.; Blum, Y.; Marisa, L.; Loncle, C.; Gayet, O.; Moutardier, V.; Turrini, O.; Giovannini, M.; Bian, B.; Bigonnet, M.; et al. Pancreatic Adenocarcinoma Therapeutic Targets Revealed by Tumor-Stroma Cross-Talk Analyses in Patient-Derived Xenograft. *Cell Rep.* **2017**, *21*, 2458–2470. [CrossRef] [PubMed]

40. Schreiber, F.S.; Deramaudt, T.B.; Brunner, T.B.; Boretti, M.I.; Gooch, K.J.; Stoffers, D.A.; Bernhard, E.J.; Rustgi, A.K. Successful Growth and Characterization of Mouse Pancreatic Ductal Cells: Functional Properties of the Ki-RAS(G12V) Oncogene. *Gastroenterology* **2004**, *127*, 250–260. [CrossRef]

41. Duconseil, P.; Gilabert, M.; Gayet, O.; Loncle, C.; Moutardier, V.; Turrini, O.; Calvo, E.; Ewald, J.; Giovannini, M.; Gasmi, M.; et al. Transcriptomic Analysis Predicts Survival and Sensitivity to Anticancer Drugs of Patients with a Pancreatic Adenocarcinoma. *Am. J. Pathol.* **2015**, *185*, 1022–1032. [CrossRef]

42. Santofimia-Castaño, P.; Rizzuti, B.; Abián, O.; Velázquez-Campoy, A.; Iovanna, J.L.; Neira, J.L. Amphipathic Helical Peptides Hamper Protein-Protein Interactions of the Intrinsically Disordered Chromatin Nuclear Protein 1 (NUPR1). *BBA Gen. Subj.* **2018**, *1862*, 1283–1295. [CrossRef] [PubMed]

43. Forli, S.; Huey, R.; Pique, M.E.; Sanner, M.F.; Goodsell, D.S.; Olson, A.J. Computational Protein-Ligand Docking and Virtual Drug Screening with the AutoDock Suite. *Nat. Protoc.* **2016**, *11*, 905–919. [CrossRef]

44. Bissantz, C.; Kuhn, B.; Stahl, M. A Medicinal Chemist's Guide to Molecular Interactions. *J. Med. Chem.* **2010**, *53*, 5061–5084. [CrossRef]
45. Ruan, H.; Sun, Q.; Zhang, W.; Liu, Y.; Lai, L. Targeting intrinsically disordered proteins at the edge of chaos. *Drug Discov. Today* **2019**, *24*, 217–227. [CrossRef]
46. Santofimia-Castaño, P.; Rizzuti, B.; Xia, Y.; Abian, O.; Peng, L.; Velázquez-Campoy, A.; Neira, J.L.; Iovanna, J. Targeting intrinsically disordered proteins involved in cancer. *Cell Mol. Life Sci.* **2020**, *77*, 1695–1707. [CrossRef]
47. Bumbak, F.; Thomas, T.; Noonan-Williams, B.J.; Vaid, T.M.; Yan, F.; Whitehead, A.R.; Bruell, S.; Kocan, M.; Tan, X.; Johnson, M.A.; et al. Conformational Changes in Tyrosine 11 of Neurotensin Are Required to Activate the Neurotensin Receptor 1. *ACS Pharmacol. Transl. Sci.* **2020**, *3*, 690–705. [CrossRef]
48. Loening, N.M.; Saravanan, S.; Jespersen, N.E.; Jara, K.; Barbar, E. Interplay of Disorder and Sequence Specificity in the Formation of Stable Dynein-Dynactin Complexes. *Biophys. J.* **2020**, *119*, 950–965. [CrossRef] [PubMed]
49. Heller, G.T.; Sormanni, P.; Vendruscolo, M. Targeting Disordered Proteins with Small Molecules Using Entropy. *Trends Biochem. Sci.* **2015**, *40*, 491–496. [CrossRef]
50. Heller, G.T.; Bonomi, M.; Vendruscolo, M. Structural Ensemble Modulation upon Small-Molecule Binding to Disordered Proteins. *J. Mol. Biol.* **2018**, *430*, 2288–2292. [CrossRef] [PubMed]
51. Crespo, A.; Fernández, A. Induced Disorder in Protein-Ligand Complexes as a Drug-Design Strategy. *Mol. Pharm.* **2008**, *5*, 430–437. [CrossRef]
52. Diehl, C.; Engström, O.; Delaine, T.; Håkansson, M.; Genheden, S.; Modig, K.; Leffler, H.; Ryde, U.; Nilsson, U.J.; Akke, M. Protein Flexibility and Conformational Entropy in Ligand Design Targeting the Carbohydrate Recognition Domain of Galectin-3. *J. Am. Chem. Soc.* **2010**, *132*, 14577–14589. [CrossRef] [PubMed]
53. Vega, S.; Kang, L.-W.; Velazquez-Campoy, A.; Kiso, Y.; Amzel, L.M.; Freire, E. A Structural and Thermodynamic Escape Mechanism from a Drug Resistant Mutation of the HIV-1 Protease. *Proteins* **2004**, *55*, 594–602. [CrossRef] [PubMed]
54. Morita, K.; He, S.; Nowak, R.P.; Wang, J.; Zimmerman, M.W.; Fu, C.; Durbin, A.D.; Martel, M.W.; Prutsch, N.; Gray, N.S.; et al. Allosteric Activators of Protein Phosphatase 2A Display Broad Antitumor Activity Mediated by Dephosphorylation of MYBL2. *Cell* **2020**, *181*, 702–715.e20. [CrossRef] [PubMed]

 *biomolecules*

Article

A Structural and Dynamic Analysis of the Partially Disordered Polymerase-Binding Domain in RSV Phosphoprotein

Christophe Cardone [1], Claire-Marie Caseau [1], Benjamin Bardiaux [2], Aurélien Thureaux [3], Marie Galloux [4], Monika Bajorek [4], Jean-François Eléouët [4], Marc Litaudon [1], François Bontems [1] and Christina Sizun [1,*]

[1] Institut de Chimie des Substances Naturelles, CNRS, Université Paris Saclay, 91190 Gif-sur-Yvette, France; cardone.christophe@gmail.com (C.C.); cmcaseau@gmail.com (C.-M.C.); marc.litaudon@cnrs.fr (M.L.); francois.bontems@cnrs.fr (F.B.)
[2] Structural Bioinformatics Unit, Department of Structural Biology and Chemistry, Institut Pasteur, CNRS UMR3528, 78015 Paris, France; bardiaux@pasteur.fr
[3] Soleil Synchrotron, SWING, 91190 Saint-Aubin, France; aurelien.thureau@synchrotron-soleil.fr
[4] Unité de Virologie et Immunologie Moléculaires, INRAE, Université Paris Saclay, 78352 Jouy-en-Josas, France; marie.galloux@inrae.fr (M.G.); monika.bajorek@inrae.fr (M.B.); jean-francois.eleouet@inrae.fr (J.-F.E.)
*   Correspondence: christina.sizun@cnrs.fr

Citation: Cardone, C.; Caseau, C.-M.; Bardiaux, B.; Thureaux, A.; Galloux, M.; Bajorek, M.; Eléouët, J.-F.; Litaudon, M.; Bontems, F.; Sizun, C. A Structural and Dynamic Analysis of the Partially Disordered Polymerase-Binding Domain in RSV Phosphoprotein. *Biomolecules* **2021**, *11*, 1225. https://doi.org/10.3390/biom11081225

Academic Editor: Simona Maria Monti

Received: 14 July 2021
Accepted: 13 August 2021
Published: 17 August 2021

Publisher's Note: MDPI stays neutral with regard to jurisdictional claims in published maps and institutional affiliations.

Abstract: The phosphoprotein P of *Mononegavirales* (*MNV*) is an essential co-factor of the viral RNA polymerase L. Its prime function is to recruit L to the ribonucleocapsid composed of the viral genome encapsidated by the nucleoprotein N. *MNV* phosphoproteins often contain a high degree of disorder. In *Pneumoviridae* phosphoproteins, the only domain with well-defined structure is a small oligomerization domain (P_{OD}). We previously characterized the differential disorder in respiratory syncytial virus (RSV) phosphoprotein by NMR. We showed that outside of RSV P_{OD}, the intrinsically disordered N-and C-terminal regions displayed a structural and dynamic diversity ranging from random coil to high helical propensity. Here we provide additional insight into the dynamic behavior of $P_{C\alpha}$, a domain that is C-terminal to P_{OD} and constitutes the RSV L-binding region together with P_{OD}. By using small phosphoprotein fragments centered on or adjacent to P_{OD}, we obtained a structural picture of the P_{OD}–$P_{C\alpha}$ region in solution, at the single residue level by NMR and at lower resolution by complementary biophysical methods. We probed P_{OD}–$P_{C\alpha}$ inter-domain contacts and showed that small molecules were able to modify the dynamics of $P_{C\alpha}$. These structural properties are fundamental to the peculiar binding mode of RSV phosphoprotein to L, where each of the four protomers binds to L in a different way.

Keywords: respiratory syncytial virus; phosphoprotein; nuclear magnetic resonance; tetramerization domain; transient secondary structure; protein dynamics

1. Introduction

A number of viral proteins are intrinsically disordered proteins (IDPs) or contain large intrinsically disordered regions (IDRs) [1–3]. Compared to globular proteins or protein domains, IDPs and IDRs sample different conformations in their free form and even in bound forms [4–6]. They can contain several binding sites, often in the form of short linear motifs of rather weak affinity, but still marked specificity [7,8]. Their flexibility is key for their functional versatility. In the case of viral IDPs, this structural adaptability compensates for the reduced number of proteins encoded by a viral genome, allowing these proteins to carry out a variety of functions at different stages of the viral cycle ranging from host cell infection, evasion of the immune system, viral genome replication to formation of new viral particles.

The RNA-dependent RNA polymerase (RdRp) complex of viruses of the *Mononegavirales* (*MNV*) order, i.e., non-segmented negative strand RNA viruses, has been widely investigated from a functional and a structural perspective [9–13]. The *MNV* RdRp complex works either in transcription or in replication mode. The *apo* form is minimally composed of two viral proteins: the large catalytic subunit (L protein) and its phosphoprotein cofactor (P protein). Its template is a ribonucleoprotein complex, formed by a genomic RNA molecule embedded in a sheath made of viral nucleoprotein (N protein). Recognition between both complexes, which leads to formation of the *holo* RdRp complex, is mediated by a direct interaction between N and P proteins [11,14–16]. The P protein not only acts as a linchpin for the *holo* RdRp complex, but also serves as an adaptor protein that tethers additional co-factors to the RdRp complex [17].

Phosphoproteins but also nucleoproteins of several *MNV* viruses, and more particularly viruses of the *Paramyxoviridae* family, were reported to contain large IDRs, either by bioinformatic analysis or by biochemical and biophysical approaches [18–20]. For example, the 525-residues N protein of measles virus (MeV) contains a 120 residues-long C-terminal tail, and MeV P protein contains 380 disordered residues out of 507 [21]. IDRs were analyzed in solution by nuclear magnetic resonance (NMR) and/or small angle X-ray scattering (SAXS), which revealed their conformational versatility [22,23] and conformational trajectories during binding events [24]. Notably, it has been shown for several MNVs, that in infected cells, viral RNA synthesis takes place in membraneless cytoplasmic inclusions with liquid-like properties, where N and P proteins are concentrated [25,26]. Formation of these viral factories was postulated to rely on a liquid-liquid phase separation mechanism, induced by weak interactions driven by the IDRs present on N and P proteins as well as RNA [1,25–27]. MeV and RSV N and P proteins also trigger liquid-liquid phase separation in vitro, and recent biochemical and biophysical characterization of N/P induced membraneless organelles confirmed the critical role of P and/or N IDRs for this mechanism in vitro and in infected cells [28,29].

Respiratory syncytial virus (RSV) is a *MNV* virus that belongs to the *Pneumoviridae* family and *Orthopneumovirus* genus. RSV is a major pathogen responsible for pneumonia in children [30] and a leading cause of hospitalization and pediatric mortality due to lower respiratory tract infections worldwide [31]. Its genome encodes 11 proteins [32]. Two of these, the phosphoprotein and the glycoprotein, were predicted to contain high disorder [33], with >50% of residues above the disorder score of 0.5 defined by the PONDR algorithm [34]. No high-resolution 3D structure is available for the RSV G protein. The 241-residue RSV P protein was shown to form stable tetramers via a central oligomerization domain (P_{OD}, residues 131–150), tetramerization being critical for its functionality [35]. RSV P_{OD} is flanked by large disordered N- and C-terminal domains (P_{NT}, residues 1–130, and P_{CT}, residues 151–241, Figure 1A) with locally high α-helical propensity [35–40].

High resolution insight into the structure of P became available when the structure of the RSV L—P complex was solved by cryo-EM [41,42]. The latter provided a structural basis for the polymerase and mRNA capping catalytic activities of RSV L [41]. It also revealed a "tentacular" binding mode of P to L, where each P protomer binds differently to L [41]. Even though the full-length P was engaged in the cryo-EM L—P complex, only the tetramerization domain and a large C-terminal part (residues 151–183, 151–186, 151–199 and 151–228, depending on the P protomers, Figure 1B) were resolved. Notably P_{NT} was missing in the cryo-EM maps.

We previously analyzed RSV phosphoprotein by NMR to acquire a more detailed structural insight into its disordered regions [17,43]. We showed that these regions displayed very different dynamic behaviors. P_{NT} is nearly fully flexible, except for two regions with low α-helical propensity. In contrast, P_{CT} is structurally heterogeneous. It contains a domain with marked α-helical propensity, which is located immediately C-terminally to P_{OD} ($P_{C\alpha}$, residues 165–205). $P_{C\alpha}$ is followed by a fully disordered C-terminal tail (P_{Ctail}, residues 210–241). By comparing our NMR results with the cryo-EM structures of the RSV L—P complex, it appears that $P_{C\alpha}$ coincides with the major part of the L-binding site.

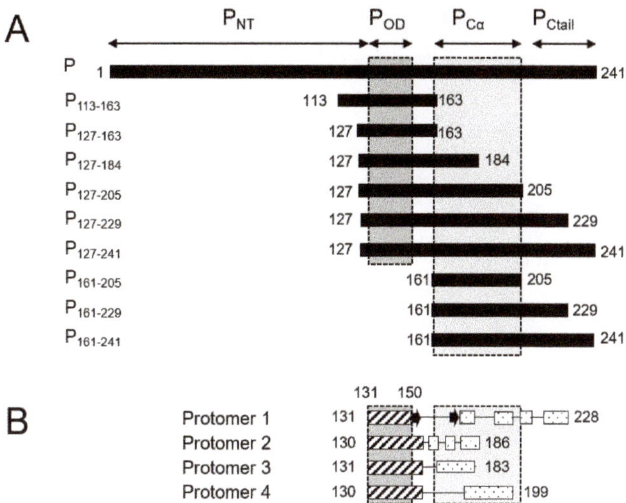

Figure 1. Schematic representation of RSV phosphoprotein (P) domains and fragments. (**A**) Boundaries of the N-terminal domain, oligomerization domain, C-terminal α-helical domain and C-terminal tail of the RSV P protein and definitions of the RSV fragments used herein. (**B**) Topology (α-helices and β-strands) of the visible parts of the four RSV P protomers in the cryo-EM structure of the RSV L—P complex (PDB 6PZK, [41]).

NMR helped rationalize the relationship between the ability of RSV phosphoprotein to recruit multiple protein partners to the RdRp complex, e.g., viral and cellular proteins required for efficient transcription [44], and several conserved small linear motifs with α-helical or β-strand secondary structure propensity in P_{NT} and P_{Ctail} [17,43]. However, the $P_{C\alpha}$ domain could not be fully characterized by NMR. This is inherent to this technique, as NMR signals are severely broadened and ultimately become undetectable when protein dynamics or exchange phenomena take place on a μs–ms timescale, precluding further analysis. This left open the question how the structure of RSV free P in solution relates to that of P in complex with L, and more particularly if the different conformations observed for the four protomers of RSV P in the L—P complex pre-exist in solution. To answer these questions, we resorted again to NMR, complemented with other biophysical techniques, such as SAXS and Dynamic Light Scattering (DLS). By using small phosphoprotein fragments centered on or adjacent to P_{OD}, we found out that NMR signals were recovered for $P_{C\alpha}$ as well as P_{OD} domains. This allowed us to establish a structural and dynamic picture of these two domains and to acquire new insight into their interactions in solution. We also showed, by analyzing binding of the small molecule garcinol, that the dynamics of $P_{C\alpha}$ can be interfered with, which opens new avenues for targeting the oligomeric structure of RSV phosphoprotein as a pharmaceutical target.

2. Materials and Methods

2.1. Expression and Purification of Proteins

RSV phosphoprotein (P) and fragments of P were expressed in *E. coli* BL21(DE3) with an N-terminal GST-tag, using plasmids derived from a pGEX vector, containing ampicillin resistance, as described previously [35]. ^{15}N- and ^{15}N^{13}C-labeled P protein samples were produced in 1 L of minimum M9 medium supplemented with 1 g·L^{-1} ^{15}NH$_4$Cl (Eurisotop, Saint-Aubin, France) and 4 g·L^{-1} unlabeled or 3 g·L^{-1} ^{13}C-labeled glucose (Eurisotop, Saint-Aubin, France). Expression was induced with 80 μg·L^{-1} IPTG overnight at 28 °C. Bacteria were harvested and disrupted (Constant Systems Ltd., Daventry, UK) in 50 mM Tris pH 7.8, 60 mM NaCl, 0.2% Triton X-100 lysis buffer supplemented with protease inhibitors (complete EDTA-free, Roche, Boulogne-Billancourt, France). After clarification by ultracentrifugation, 2 mL of glutathione Sepharose resin beads (GE Healthcare, Buc, France) were added to the supernatant, and the mixture was incubated overnight at 4 °C. The resin beads were then washed with thrombin cleavage buffer (20 mM Tris pH 8.4, 150 mM NaCl, 2.5 mM $CaCl_2$) before addition of 5 units of biotinylated thrombin (Novagen, Merck Millipore, Guyancourt, France). After incubation for 16 h at 4 °C, the cleaved protein was eluted and thrombin was captured with streptavidin resin according to the manufacturer's instructions (Novagen, Merck Millipore, Guyancourt, France). In a final step, samples were dialyzed into NMR buffer (20 mM sodium phosphate, pH 6.5, 100 mM NaCl) and concentrated to 50–300 µM on Amicon Ultra centrifugal filters with 10 kDa cut-off (Merck Millipore, Guyancourt, France). Some samples were exchanged into D_2O buffer (prepared by reconstituting lyophilized NMR buffer into 100% D_2O (Eurisotop, Saint-Aubin, France) by passing 4 times through a desalting column (Biospin, Bio-Rad, Les Ulis, France) equilibrated in this buffer. All samples were analyzed by SDS-PAGE. Protein concentrations were determined from Bradford or BCA assays and by measuring the absorbance at 280 nm for proteins containing tyrosine residues.

2.2. Nuclear Magnetic Resonance (NMR) Spectroscopy and Chemical Shift Assignments

2D and 3D NMR experiments for backbone assignment were carried out on Bruker Avance III spectrometers at magnetic fields of 14.1, 16.4 or 18.8 T (600, 700 or 800 MHz ^1H Larmor frequencies, respectively), equipped with cryogenic TCI probes. Samples contained 7% D_2O to lock the magnetic field. ^1H and ^{13}C chemical shifts were referenced to 4,4-dimethyl-4-silapentane-1-sulfonic acid (DSS). Spectra were processed with Topspin 4.0 (Bruker Biospin, Wissembourg, France) and analyzed with CCPNMR 2.5 [45] software.

Sequential backbone assignment of $^{15}N^{13}C$ labeled proteins was carried out using BEST-TROSY versions of triple resonance 3D HNCO, HNCA, HN(CO)CA, HNCACB, HN(CO)CACB experiments [46] and standard versions of 3D CBCA(CO)NH and HBHA(CO)NH experiments. Amide $^1H/^{15}N$ and $^1H_\alpha$ assignments of ^{15}N labeled proteins was carried out by recording 3D ^{15}N-separated DIPSI-HSQC and NOESY-HSQC spectra. A 3D (H)CCH-TOCSY spectrum was acquired at 800 MHz on a $P_{127-163}$ sample in 100% D_2O buffer for sidechain assignment.

2.3. High-Pressure NMR Measurements

Steady-state high pressure NMR measurements were carried out in a zirconia ceramic NMR tube (Daedalus Innovations, Aston, PA, USA) filled with 350 µL of 400 µM $^{15}N^{13}C$ labeled $P_{127-205}$ and 20 µL D_2O. Pressure was generated with an X-treme 60 syringe pump (Daedalus Innovations) filled with mineral oil. The temperature was set to 313 K. ^1H-^{15}N HSQC spectra were acquired at variable pressure between 1 bar and 2.5 kbar, with 100–500 bar increments. 3D HNCA spectra used for chemical shift assignment were acquired at 500 and 2500 bar.

2.4. NMR Structure Determination of $P_{127-163}$

Distance restraints to calculate the NMR structure of $P_{127-163}$ were derived from NOESY experiments performed on two samples: a uniformly $^{15}N^{13}C$-labeled sample and a sample with mixed labeling. The latter was reconstituted from an equimolar mixture of unlabeled and $^{15}N^{13}C$-labeled $P_{127-163}$ incubated overnight at 4 °C with 6 M guanidinium chloride. The sample was then dialyzed at 4 °C for two hours on a membrane with a 3.5 kDa cut-off, successively into NMR buffer supplemented with 8 M urea (200 mL), NMR buffer with 2 M urea (3 × 500 mL) and NMR buffer (3 × 1 L).

Intra- and inter-protomer distance restraints were obtained by measuring the signal heights in two 3D NOESY-HSQC experiments with 80 ms mixing time acquired on the $P_{127-163}$ sample with uniform labeling: a ^{15}N-separated NOESY-HSQC in H_2O buffer and a ^{13}C-separated NOESY-HSQC in D_2O buffer. They were treated as ambiguous restraints. Unambiguous inter-protomer distances were obtained from a ^{13}C-separated NOESY-HSQC with a ^{13}C-filter in the t1 dimension, measured on the mixed sample in D_2O buffer. All NOESY data were acquired at a magnetic field of 22.3 T (950 MHz ^{1}H Larmor frequency). Dihedral angle restraints were obtained from backbone chemical shifts by using the TalosN software [47]. 3D structure calculation of tetrameric $P_{127-163}$ was performed with the ARIA 2.3.2 software [48] imposing a C4 symmetry [49].

NMR restraints and structural statistics for RSV $P_{127-163}$ were generated with ARIA and the Protein Structure validation suite V1.5 (http://psvs.nesg.org/, http://rtti7.uv.es/, accessed on 15 April 2020).

2.5. Paramagnetic Spin Labelling and Paramagnetic Relaxation Enhancement (PRE) Measurements

Two serine to cysteine mutants of $P_{127-205}$, S143C and S156C, were constructed using the QuickChange mutagenesis kit (Stratagene, Agilent, Les Ulis, France). The mutations were verified by sequencing. Mutant proteins were expressed and purified such as wild type proteins. After purification, protein samples were treated with 10 mM DTT at room temperature for 2 h to ensure that cysteine residues were in a reduced state. DTT was then removed by passing twice through a Biospin desalting column (Biospin, Biorad, Les Ulis, France) equilibrated with 50 mM Tris, pH 8.0, 200 mM NaCl. The cysteine mutants (200 µL at 300 µM final concentration) were reacted overnight at 15 °C in the dark with 10 molar equivalents of 3-(2-iodoacetamido)-PROXYL radical (IAP, Sigma-Aldrich, Saint-Quentin-Fallavier, France) taken from a 45 mM stock solution in ethanol. Unreacted IAP was removed by passing the samples three times through a desalting column equilibrated in the final NMR buffer. After data acquisition with the paramagnetic IAP spin label, the sample was treated with 10 molar equivalents of ascorbic acid (Sigma-Aldrich, Saint-Quentin-Fallavier, France) at pH 6.5 overnight in the dark to generate a diamagnetic spin label.

Paramagnetic relaxation enhancements (PREs) were determined for each residue as the ratio of peak intensities measured in the ^{1}H-^{15}N HSQC spectra of samples with spin labels in the paramagnetic versus diamagnetic state (I_{para}/I_{dia}). NMR data were acquired at a magnetic field of 14.1 T and a temperature of 313 K. To model the PRE profiles, 1000 protein conformers were simulated with the Flexible-meccano software [50], assuming 100% α-helical propensity for residues 130–156 and using the α-helical propensities determined from backbone chemical shifts with TalosN [47] for residues 157–205. The segments 143–156 (containing the labeled positions) and 160–199 (where strong PREs were measured) were defined as contact regions. PREs were simulated with no contact and contact distances of 7, 10 and 15 Å.

2.6. Small Angle X-ray Scattering (SAXS)

SAXS experiments were performed at the SWING beamline at the SOLEIL synchrotron (Saint-Aubin, France), using an EigerX 4M detector at a distance of 2 m. The temperature was set to 293 K. $P_{127-163}$ and $P_{127-205}$ solutions were prepared in 20 mM Na phosphate pH 6.5, 100 mM NaCl buffer at final concentrations of 200 µM and 325 µM, respectively. Protein samples (50 µL) were injected onto a size exclusion column (BioSec3-300, Agilent, Buc, France) and eluted directly into the SAXS flow-through capillary cell at a flow rate of 0.3 mL/min, as previously described [51]. 510 SAXS frames (exposure time: 990 ms, dead time: 10 ms) were collected continuously during elution. Data reduction to absolute units, buffer subtraction and averaging of identical frames corresponding to the elution peak (10 and 13 frames for $P_{127-163}$ and $P_{127-205}$, respectively) were conducted with the SWING in-house software FOXTROT [52] and US-SOMO [53]. I(0) and Rg values were determined with the PRIMUS software [54]. Normalized Kratky curves were obtained by plotting $(q*Rg)^2*I(q)/I(0)$ as a function of $q*Rg$, where q is the scattering vector amplitude. The theoretical normalized Kratky plot for a globular protein was calculated using the expression $(q*Rg)^2*exp(-(q*Rg)^2/3)$. The SAXS data intensity scattering curve of $P_{127-163}$ was fitted on the FoXs server [55] using the NMR structure of $P_{127-163}$ and declaring the 9 N- and 15 C-terminal residues as flexible.

2.7. Interaction Experiments with Garcinol

Stock solutions of garcinol (CAS number 78824-30-3) were prepared in DMSO (3 mM) or in ethanol (30 mM) and added to 150 µM $P_{161-241}$ and 300 µM $P_{127-205}$, up to a molar ratio of 3:1 and 1:1, respectively. The mixtures were analyzed by NMR at a Larmor frequency of 600 MHz ^{1}H and at a temperature of 293 K for $P_{161-241}$, and at 700 MHz and 313 K for $P_{127-205}$.

2.8. Dynamic Light Scattering (DLS)

Dynamic light scattering measurements were carried out on a Zetasizer Nano S instrument (Malvern, Orsay, France) at a backscatter angle of 173° and a temperature of 298 K. $P_{127-205}$ samples, used for NMR before, were centrifuged before being transferred into disposable 40 µL cuvettes (Brand Z637092, Wertheim, Germany). Size distribution analysis was performed within the Zetasizer software, by setting the refractive index to 1.450 and the viscosity to 0.8872 cP.

2.9. Illustrations

Figures were prepared using Pymol [56].

2.10. Data Depositions

RSV $P_{127-163}$ chemical shifts were deposited at the BMRB with accession number 34513. Coordinates and restraints for the NMR structure of RSV $P_{127-163}$ were deposited at the Protein Data Bank under accession code 6YP5. SAXS data for $P_{127-163}$ were deposited at the SASBDB with accession code SASDJD2.

3. Results

3.1. Signal Broadening in NMR Spectra of RSV Phosphoprotein Fragments Reveals Regions with Partial to High Structural Order

2D ^{1}H-^{15}N HSQC spectra provide structural and dynamic fingerprints of ^{15}N-labeled proteins at the single residue level, as each amide pair accounts for an individual signal, identified by its ^{1}H and ^{15}N chemical shifts. As the chemical shift of an atomic nucleus is sensitive to its electronic environment, it is a highly sensitive local probe for structural complexity. As IDPs and IDRs lack defined 2D or 3D structure, they display a narrow amide ^{1}H chemical shift dispersion. Moreover, IDP/IDR signals are sharp due to rapid motions of the polypeptide chain. In contrast, the amide ^{1}H chemical shift dispersion of well-structured protein regions is wider. Signal linewidths depend on nuclear relaxation, which depends on molecular tumbling. Hence, as a general rule, linewidths increase concomitantly with molecular size, but conformational and chemical exchange contribute to additional line broadening.

We previously measured ^{1}H-^{15}N HSQC spectra for the RSV full length P. One third of the signals was missing due to severe line broadening [43]. We identified the missing signals as those of the oligomerization domain P_{OD} and of the $P_{C\alpha}$ domain. When we measured ^{1}H-^{15}N HSQC spectra of RSV P N- and C-terminal fragments containing P_{OD}, the same signals were missing [43]. $P_{C\alpha}$ signals were only detected in monomeric fragments without P_{OD}. This raised the question if the structural and dynamic information obtained from these monomeric fragments by chemical shift and nuclear relaxation analysis was relevant for the tetrameric protein.

We therefore sought new conditions to recover P_{OD} and $P_{C\alpha}$ signals for tetrameric fragments in amide-detected NMR experiments. We produced eight ^{15}N-labeled fragments of RSV phosphoprotein of various sizes (Figure 1A). The $P_{127-163}$ fragment roughly corresponds to P_{OD}. The $P_{113-163}$ fragment has an N-terminal extension with high sequence conservation within the *Orthopneumovirus* genus. $P_{127-184}$ has a C-terminal extension that is conserved throughout the *Pneumoviridae* family, whereas the longer $P_{127-205}$ fragment is only well conserved within *Orthopneumoviruses* [17]. $P_{127-229}$ contains the full L-binding site revealed in the cryo-EM RSV L—P complex structure [41,42]. $P_{127-241}$, which consists of P_{OD} and the entire C-terminal domain, was investigated previously and reproduced the behavior of full-length RSV phosphoprotein [43]. In addition, we produced two C-terminal fragments without P_{OD}, $P_{161-205}$ and $P_{161-229}$, for comparison.

We measured their ^{1}H-^{15}N HSQC spectra under identical buffer and temperature conditions. We first recorded NMR data at low temperature, i.e., at 288 K (Figure 2A). This condition is more favorable to observe amide signals that are not or only loosely involved in hydrogen bonds, i.e., signals from IDRs. The signals were sharp and not well dispersed, which pointed at a disordered state of the protein fragments. For most fragments, the number of observed signals was less than expected. For $P_{113-163}$, only a set of ~20 sharp signals was observed, corresponding to residues 113–130 and 158–163. The signals of residues 131–157 were 5 to 10-fold less intense than the set of sharp signals; they became visible by lowering the contour levels close to the noise level. For $P_{127-184}$, ~25 sharp signals assigned to residues 157–184 were observed, but the signals from P_{OD} were missing. For $P_{127-205}$ only a few flexible residues of the N-and C-terminal extremities were visible. For $P_{127-229}$ and $P_{127-241}$, ~30 and ~40 resonances were detected; they were assigned to residues 204–229 and 203–241, respectively, and correspond to P_{Ctail}. In contrast, no signal was missing for the two constructs without P_{OD}, $P_{161-229}$ and $P_{161-205}$. The results are in overall agreement with previous results obtained with longer fragments including P_{1-127}, P_{1-163} and $P_{161-241}$ [43]. Comparison between $P_{127-229}$ and $P_{161-229}$ (superimposed in Figure 2A) indicates that the chemical shifts of P_{Ctail} are independent of the presence of P_{OD}, confirming that this part of RSV P is a fully disordered IDR.

In contrast to P_{Ctail} signals, $P_{C\alpha}$ signals were broadened out in nearly all P_{OD}–containing fragments, with the notable exception of $P_{127-184}$. In the latter case, the sharpness of $P_{C\alpha}$ signals indicates that residues 157–184 were disordered and flexible, as a consequence of truncation within the $P_{C\alpha}$ region. Hence, $P_{127-184}$ is not representative of the full-length protein, contrary to $P_{127-205}$.

The signals of P_{OD} were also broadened out, except in $P_{127-163}$, a P_{OD}–containing fragment with very short N- or C-terminal extensions. The amide ^{1}H chemical shifts of $P_{127-163}$ were only little more dispersed than those of fully disordered regions, but this remains compatible with helical secondary structure. Linewidths in $P_{127-163}$ were broader than for RSV P IDRs, which is indicative of the presence of a stably folded domain. Together these results suggest that extensions longer than 10 residues induce motions on the μs–ms time scale, presumably shearing motions in the coiled coil P_{OD}, resulting in signal broadening.

In addition to NMR data at a temperature of 288 K, we acquired data at 313 K to take advantage of signal narrowing for folded protein signals due to less efficient nuclear relaxation at higher temperature. We focused on the three P_{OD}–containing fragments with C-terminal extensions: $P_{127-163}$, $P_{127-184}$ and $P_{127-205}$ (Figure 2B). At 313K, the P_{OD} signals (residues 130–154) were indeed observable and superimposed well between the three fragments, indicating that P_{OD} adopted the same structure. The $P_{C\alpha}$ signals of $P_{127-184}$ and $P_{127-205}$ were also observed, but did not superimpose, confirming that the residues 160–184 adopt different conformations in both fragments. P_{OD} signals were less intense and broader than those of the C-terminal extensions in the $P_{127-184}$ and $P_{127-205}$ fragments, which indicates that $P_{C\alpha}$ is not stably associated, neither in a coiled coil conformation such as P_{OD}, nor directly bound to P_{OD}.

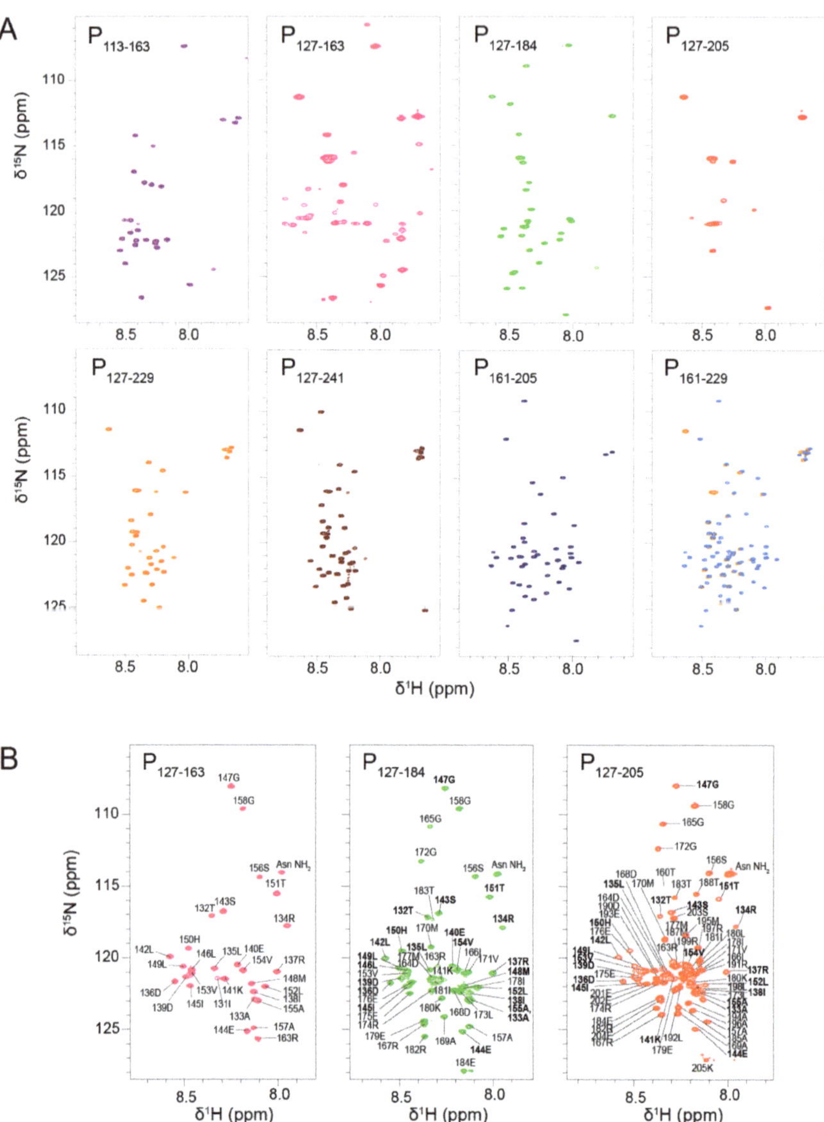

Figure 2. NMR fingerprints of RSV phosphoprotein fragments. (**A**) 2D ^{1}H-^{15}N HSQC spectra of eight ^{15}N-labeled RSV P fragments were recorded at low temperature (288 K) in 20 mM phosphate pH 6.5, 100 mM NaCl buffer and at 150–300 μM concentrations. These fragments either contain the oligomerization domain (P$_{OD}$) with an N-terminal extension (P$_{113–163}$), only P$_{OD}$ (P$_{127–163}$), P$_{OD}$ with C-terminal extensions of different lengths (P$_{127–184}$, P$_{127–205}$, P$_{127–229}$ and P$_{127–241}$) or only C-terminal extensions (P$_{161–205}$ and P$_{161–229}$). NMR data were measured at 600 MHz ^{1}H Larmor frequency for all fragments, except P$_{113–163}$ that was measured at 700 MHz. Each fragment was assigned a color: P$_{113–163}$ (purple), P$_{127–163}$ (pink), P$_{127–184}$ (green), P$_{127–205}$ (red), P$_{127–229}$ (orange), P$_{127–241}$ (maroon), P$_{161–205}$ (marine), and P$_{161–229}$ (medium blue). The spectrum of P$_{127–229}$ (orange) was superimposed onto that of P$_{161–229}$ (blue). Spectra were plotted with comparable contour levels. (**B**) 2D ^{1}H-^{15}N HSQC spectra of three ^{15}N-labeled RSV P fragments were measured at high temperature (313 K) on a 600 MHz NMR spectrometer. Amide signals are annotated with residue number and amino acid type. P$_{OD}$ residues are highlighted in bold letters.

3.2. Secondary Structure of RSV P_{OD} and $P_{C\alpha}$ by NMR

To gain more insight into the structures of $P_{127-184}$ and $P_{127-205}$, we analyzed secondary structure propensities for each residue based on backbone chemical shifts at 313 K. Chemical shifts were determined by sequential assignment from 3D triple resonance and ^{15}N-separated NOESY-HSQC experiments. We used the TalosN server, which predicts α-helix, β-strand and random coil propensities [47]. The input consisted of ^{13}Cα, ^{13}Cβ, ^{13}CO and ^{1}Hα chemical shift values. We determined nearly 100% α-helical propensity for the P_{OD} part in both fragments and significant α-helical propensity in the $P_{C\alpha}$ part (Figure 3A). The short linker region between P_{OD} and $P_{C\alpha}$, around residue T160, has nearly 100% random coil propensity.

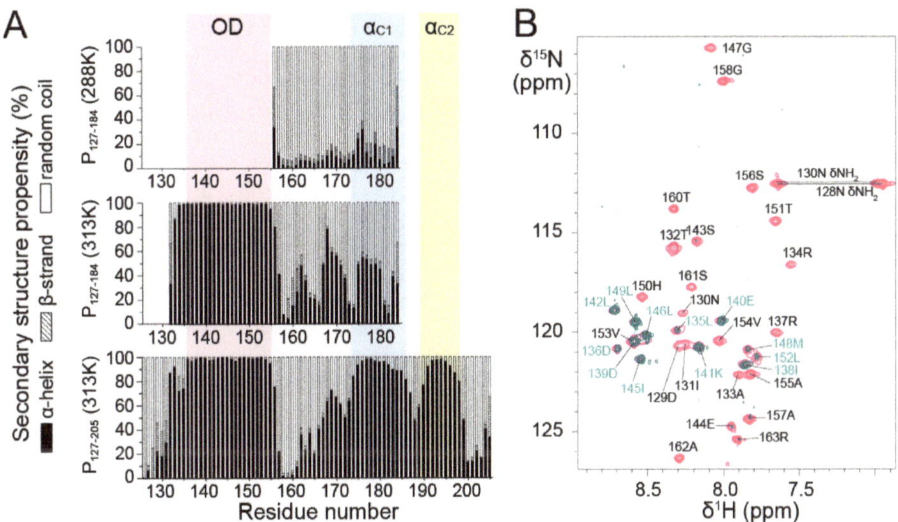

Figure 3. NMR data and secondary structure propensities of RSV phosphoprotein fragments containing P_{OD} and C-terminal extensions of variable lengths. (**A**) Secondary structure propensities for α-helix, β-strand and random coil were determined from backbone chemical shift by using the TalosN server [47]. ^{13}Cα, ^{13}CO, ^{13}Cβ and ^{1}Hα values were used as an input for $P_{127-184}$ and $P_{127-205}$ at 313 K. Only ^{13}Cα values were available for $P_{127-184}$ at 288 K. (**B**) Amide ^{1}H/^{2}H exchange experiment by NMR. 2D ^{1}H-^{15}N HSQC spectra of ^{15}N-labeled $P_{127-163}$ before (pink) and after exchange into 100% ^{2}H$_2$O buffer (teal) were overlaid. Measurements were carried out at a temperature of 298 K on an 800 MHz NMR spectrometer. Assignments for amides that are protected from water exchange are indicated in cyan.

$P_{C\alpha}$ residues displayed less α-helical (<50%) and more random coil propensity in $P_{127-184}$ than in $P_{127-205}$ at 313 K. This suggests that the 185–205 region promotes upstream α-helix formation in the $P_{C\alpha}$ domain. The increased disorder of the truncated $P_{C\alpha}$ domain in $P_{127-184}$ was also found at lower temperature, since at 288 K secondary structure propensities derived from ^{13}Cα chemical shifts were mostly random coil for residues 155–184, with <20% α-helical propensity (Figure 3A).

Even though the NMR data of $P_{127-205}$ confirm that $P_{C\alpha}$ has high α-helical propensity, they also show that $P_{C\alpha}$ does not form a long stable helix such as P_{OD}. In $P_{127-205}$, $P_{C\alpha}$ consists of two or three regions with α-helical propensities >50%, but still <100% (Figure 3A). Two regions, termed α_{C1} (residues 174–185) and α_{C2} (residues 189–198), coincide with two transient C-terminal α-helices previously determined from the monomeric fragments $P_{161-241}$ and $P_{1-121\,+\,161-241}$ by solution NMR at 288 K [43].

In a ^{15}N NOESY-HSQC experiment performed with $P_{127-205}$, the amide protons of P_{OD} displayed (i, i+1) Nuclear Overhauser Effects (NOEs) indicative of α-helices, but no NOEs with water, showing that they were protected against water exchange and engaged in stably formed α-helices. The $P_{C\alpha}$ region lacked NOEs indicative of stable helix formation. This comforts the hypothesis that P_{OD} and $P_{C\alpha}$ form independent structural entities, with different tumbling properties and that the α-helices in $P_{C\alpha}$ remain transient and dissociated.

It is noteworthy that α_{C1} corresponds to a region that adopts an α-helical conformation in all four P protomers in the RSV L—P complex, whereas α_{C2} is stabilized into α-helix for only two protomers [41,42]. The 163–173 region, which displays less but still significant α-helical propensity in $P_{127–205}$, is stabilized into an α-helix in only one protomer in the RSV L—P complex. In this complex, one P protomer forms two additional C-terminal helices spanning residues 205–211 and 215–233. These do not display any significant α-helical propensity in the free P [43], suggesting that they only fold on binding. Our results thus show that the different conformations observed for the four protomers of RSV P in the L—P complex partly pre-exist in solution. These helices are on the most ordered side of the spectrum of IDRs/IDPs, but remain transient in solution.

Furthermore, we found that residues 135–154, located in the P_{OD} region of $P_{127–205}$, have ~100% α-helical propensity. The coiled coil core domain in solution thus appears to be slightly shifted with respect to the tetrameric core of P in the cryo-EM L—P structure. The latter comprises only residues 131–150, but three protomers remain associated up to residue S156, which is closer to the boundaries found in solution. To test the hypothesis of helix fraying in P_{OD}, we performed amide $^1H/^2H$ exchange with the $P_{127–163}$ fragment. $^1H/^2H$ exchange results in vanishing ^1H-^{15}N HSQC signals for amides that are not involved in strong hydrogen bonds. As expected, the N- and C-terminal ends of $P_{127–163}$, spanning residues D129–R134 and V154–R163, respectively, were fully $^1H/^2H$ exchanged (Figure 3B). Only a few amides located in the middle of the sequence (D136, D139–L142, I145–L146 and L149) were completely exchange-protected. L135, I138, M148, L152 and V153 amides were partially protected, which suggests that the helices start fraying at both ends. Surprisingly, several other central residues (R137, S143–E144, G147, H150 and T151) were nearly completely exchanged, suggesting that the P_{OD} helices are not perfectly straight and contain kinks that account for solvent accessibility in the 137–151 region.

3.3. Probing the Stability of the RSV P_{OD}–$P_{C\alpha}$ Region by High Pressure NMR

High pressure NMR allows to locally probe protein unfolding due to mechanical compression [57]. A general rule is that with increasing pressure, protein conformational equilibria are shifted towards states with smaller partial volume and increasing conformational disorder [58]. Even if high pressure NMR best applies to globular proteins with internal cavities, leading to large partial volume changes, we nevertheless tested this technique on the RSV fragment $P_{127–205}$, i.e., the P_{OD}–$P_{C\alpha}$ region, which already contains a certain amount of disorder. This approach was driven by two considerations. First, solvation plays a role in the pressure response of proteins [58,59], which applies to any kind of protein. Second, protein chemical shifts have been reported to be highly sensitive to pressure [58]. Amide ^1H and ^{15}N chemical shifts are expected to display a linear pressure dependence for rigid proteins and to deviate from linearity for proteins in multiple conformational states [60].

We measured ^1H-^{15}N HSQC spectra of $P_{127–205}$ at a temperature of 313 K, where all signals were observable, and increased pressure from 1 to 2500 bar in 100–500 bar steps. We observed amide chemical shift perturbations (Figure 4A). To correctly re-assign amide chemical shifts at each pressure step, we checked the sequential assignment in two 3D HNCA experiments acquired at 500 and 2500 bar. In addition to chemical shift perturbations, we also observed intensity perturbations. For several residues, e.g., R137 and G147 located in P_{OD}, the intensity decreased; for other residues, e.g., T160 and G172 located in $P_{C\alpha}$, the intensity increased (Figure 4A). This was a first hint that P_{OD} and $P_{C\alpha}$ could be discriminated by high pressure NMR. These intensity variations suggest that conformational equilibria take place and that they shift with variable pressure. In the case of P_{OD}, the coiled coil conformation may be destabilized. In the case of $P_{C\alpha}$, which contains transient helices, the conformational equilibria that exist at ambient pressure are also shifted, likely towards a more disordered state. However, $^{13}C\alpha$ chemical shifts measured at 500 and 2500 bar did not show significant perturbations as compared to ambient pressure, indicating that increasing pressure did not significantly affect secondary structure propensities in $P_{C\alpha}$.

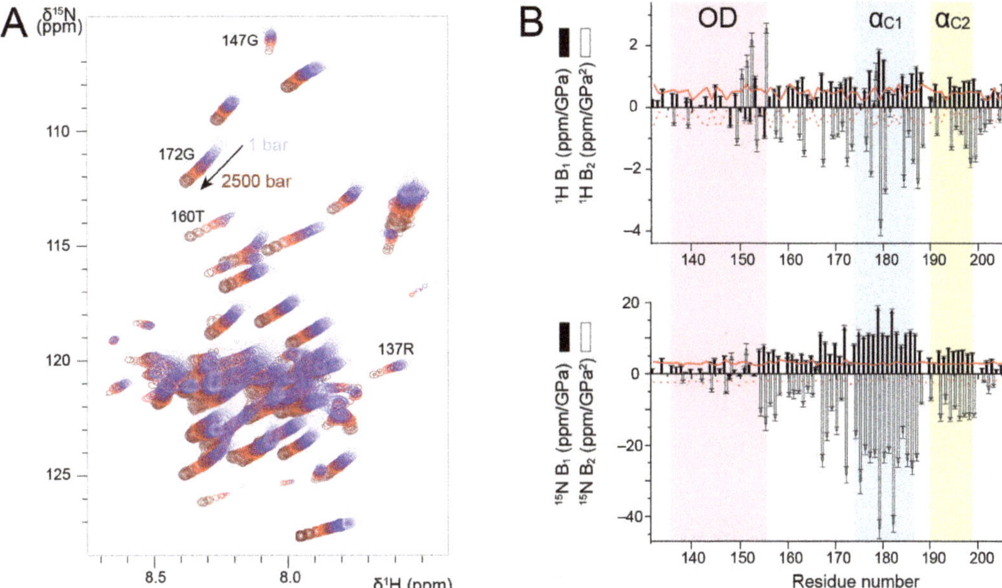

Figure 4. Discrimination between stably associated and isolated helices in RSV $P_{127-205}$ by variable pressure NMR experiments. (**A**) Overlay of ^1H-^{15}N HSQC spectra of $P_{127-205}$ acquired by increasing pressure from 1 to 2500 bar. NMR data were recorded at 600 MHz ^1H Larmor frequency. The temperature was 313 K. (**B**) First and second order pressure coefficients, B_1 and B_2, were determined by fitting ^1H and ^{15}N chemical shift variations as a function of pressure to a second-order polynomial according to $\delta(P) - \delta(P_0) = B_1 \times (P - P_0) + B_2 \times (P - P_0)^2$, and plotted with bar diagrams. Reference values for random coil conformation (taken from [60]) were plotted as red lines (B_1, solid line) and (B_2, dotted line).

We analyzed amide ^1H and ^{15}N chemical shifts into more details by fitting the pressure dependency with a second-order polynomial. We extracted linear (B_1) and quadratic (B_2) pressure coefficients (Figure 4B) and compared them to reference values tabulated for each amino acid type in random coil conformation [60]. ^1H B_1 values were slightly higher for $P_{C\alpha}$ than for P_{OD}, but they were all close to the reference values. ^1H B_2 values deviated more from the reference values, and they were more negative for $P_{C\alpha}$ than for P_{OD}. For residues in the vicinity of H150, the deviations observed for ^1H B_2 may be accounted for by pressure induced pH variations [61]. ^1H pressure coefficients thus do not allow unambiguously distinguishing between P_{OD} and $P_{C\alpha}$. ^{15}N B_1 and B_2 values draw a clearer picture of the structural difference between these domains. For P_{OD}, ^{15}N B_1 and B_2 values were small and close to the reference values. In the $P_{C\alpha}$ domain, both B_1 and B_2 significantly deviated from the reference values. The transient helices α_{C1} and α_{C2} displayed the largest deviations (Figure 4B). Taken together, our results suggest that ^{15}N B_1 and B_2 pressure coefficients allow discriminating between the isolated $P_{C\alpha}$ helices and the coiled coil P_{OD}, and that transient helices are characterized by high ^{15}N B_1 and B_2 values.

3.4. NMR Structure of P_{OD} and Spatial Excursions of $P_{C\alpha}$

To acquire further insight into the solution structure of P_{OD}, we solved the structure of the $P_{127-163}$ fragment by NMR, using dihedral angle and Nuclear Overhauser Effect

(NOE) distance restraints. We assigned >97% of sidechain ^{1}H chemical shifts of $P_{127-163}$. The main difficulty was to distinguish between intra- and inter-protomer NOEs due to the oligomeric nature of P_{OD}. To overcome this difficulty, we produced three $P_{127-163}$ samples: a first fully $^{13}C^{15}N$ labeled sample, a second unlabeled sample and a third mixed unlabeled and $^{13}C^{15}N$ labeled sample. To produce the latter, we denatured a 50:50 mixture of unlabeled and $^{13}C^{15}N$ labeled $P_{127-163}$ with 6 M guanidinium chloride. After renaturation in a buffer without chaotropic agent, $P_{127-163}$ reassembled with the same protomer arrangement, as judged from ^{1}H and ^{13}C chemical shifts that remained the same after the denaturation/renaturation treatment. We measured distance restraints for a single protomer from conventional NOESY experiments using the fully $^{13}C^{15}N$ labeled $P_{127-163}$ sample. We measured and assigned inter-protomer restraints from an edited/filtered NOESY experiment using the mixed labeled $P_{127-163}$ sample. Of note, the detected NOEs were compatible with a parallel arrangement of the tetramer.

As $P_{127-163}$ displayed a single set of chemical shifts, and since P was proposed to form tetramers [35], we assumed a C4 symmetry for structure calculations with the ARIA software [48], using this symmetry (Figure 5A). The structural statistics are given in Supplementary Table S1. The α-helical core domain spans residues N128–A157. The six most C-terminal residues are disordered. An 8-amino acid long N-terminal stretch, which is a leftover from a glycine and serine linker between the GST-tag and the protein sequence after thrombin cleavage, is also disordered. Superimposition of the NMR structure with the cryo-EM structure of P_{OD} in Figure 5B shows that the arrangement of the coiled coil domain is comparable to that in the RSV P—L complex [41,42]. The helices in the NMR structure appear to be more bent, with a small kink at position S143, which reminds of the increased water accessibility observed at this position. The helix length of the NMR structure corresponds to that of the longest helix of the coiled coil domain in the cryo-EM structure (Figure 5B). This suggests that the conformation of the C-terminal end of each bound protomer in the P—L complex is determined by its interactions with L. We calculated the electrostatic surface potential of $P_{127-163}$ using the Delphi software [62] (Figure 5A). The surface of the coiled coil domain at the N-terminal side is charged, both negatively and positively, up to residue E144. The C-terminal side, starting at residue I145, is completely hydrophobic. This confers different interaction properties, notably within the RSV P—L complex. In the cryo-EM P—L complex structure, the N-terminal side of P_{OD} sticks out, whereas the C-terminal side is associated to L.

To confirm the oligomerization state of $P_{127-163}$ in solution we performed SAXS measurements. A Guinier plot afforded a radius of gyration (R_g) of 21.2 Å, which is roughly in agreement with a coiled coil domain length of ~50 Å measured directly from the NMR structure. We fitted the scattering curve with the structure of $P_{127-163}$ using the FoXs server [55]. Even though the fit was not perfect, a tetrameric structure best fitted the curve, as compared to monomeric, dimeric or trimeric structures (Figure 5C).

Normalized Kratky plots were calculated using the R_g and I(0) values obtained with the PRIMUS software [54]: 21.2 Å and 0.0447 for $P_{127-163}$; 36.8 Å and 0.0749 for $P_{127-205}$. The bell-shape of the normalized Kratky plot for $P_{127-163}$ for q.Rg < 6 and the position of the bell maximum at q.Rg $\approx \sqrt{3}$ [63] indicate that $P_{127-163}$ is mostly globular (Figure 5C). The deviations from the theoretical curve for a globular protein may be ascribed to the disordered N- and C-terminal residues. The deviation from the bell-shape is larger for the $P_{127-205}$ fragment (Figure 5C). Our SAXS data thus confirm that the P_{OD}–$P_{C\alpha}$ region is not globular and comfort the hypothesis that the $P_{C\alpha}$ helices are not stably associated in solution.

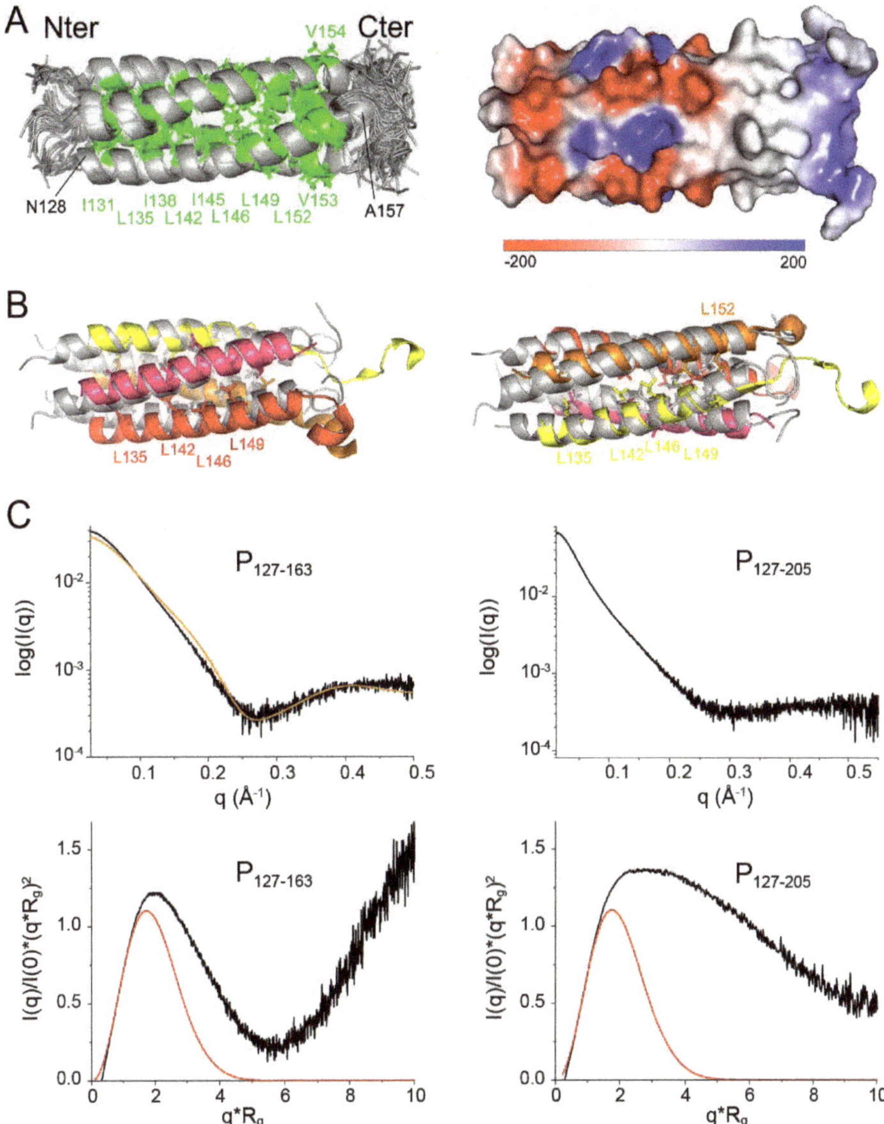

Figure 5. Solution structure of the oligomerization domain of RSV P and dynamics of its C-terminal extensions. (**A**) NMR ensemble structure of RSV $P_{127-163}$. 20 conformer structures calculated with the ARIA software [48] were structurally aligned. Backbones are represented in cartoon and the three hydrophobic residues Ile, Leu and Val in green sticks. The electrostatic surface potential was calculated using the Delphi software [62]. The scale for the ensemble of 20 structures ranges from -200 k_BT (red) to $+200$ k_BT (blue). (**B**) Superimposition of the NMR structure of RSV $P_{127-163}$ (conformer 1, in grey) onto the cryo-EM structure of the same sequence in the P—L complex (PDB 6pzk). Each protomer has a different color: yellow, orange, magenta and red. Leucine side chains are shown in sticks. (**C**) SAXS data were measured for $P_{127-163}$ and $P_{127-205}$. The scattering intensity curves ($\log(I) = f(q)$), where q is the scattering vector amplitude) are shown. The scattering intensity curve corresponding to the NMR structure of $P_{127-163}$ (orange line) was calculated with the FoXs server [55] and superimposed onto the experimental data. Normalized Kratky plots ($(q*R_g)^2*I(q)/I(0) = f(q*R_g)$) were calculated using the R_g and $I(0)$ values obtained with the PRIMUS software [54]: 21.2 Å and 0.0447 for $P_{127-163}$; 36.8 Å and 0.0749 for $P_{127-205}$. The theoretical normalized Kratky plot for a globular protein is plotted with a red line.

3.5. Paramagnetic Relaxation Enhancement Reports on Transient Contacts between the $P_{C\alpha}$ and P_{OD} Domains

To further investigate the relative flexibility of the $P_{C\alpha}$ domain with respect to P_{OD}, we measured paramagnetic relaxation enhancements by NMR [64]. We previously applied this technique to full-length phosphoprotein to detect transient contacts within the tetrameric protein. We found transient contacts within the disordered N-terminal domain of the phosphoprotein, but also between P_{NT} and P_{OD} [43]. Here we investigated the $P_{127-205}$ fragment, where the P_{OD}–$P_{C\alpha}$ region can be observed, in contrast to the full-length phosphoprotein.

We produced two $P_{127-205}$ samples with a paramagnetic nitroxide spin label (3-(2-iodoacetamido)-PROXYL radical, IAP) at positions 143 and 156, after mutating serine residues 143 and 156 into cysteines. S143 is located inside the coiled coil domain and its sidechain points outward. S156 is located at the C-terminal end of the coiled coil and oriented inward. We first assessed the influence of the mutations on the structure by measuring ^{1}H-^{15}N HSQC spectra. The spectra of both S143C and S156C mutants displayed chemical shift perturbations up to six residues from the mutated site (Figure 6A). Otherwise, the spectra overlaid well with the WT spectrum, indicating that the mutations did not disrupt the overall structure of $P_{127-205}$. This was confirmed after treatment of the IAP labeled samples with ascorbate, as the two spectra with diamagnetic spin labels were similar, except for the mutated sites and adjacent regions (Figure 6B). The signals of the diamagnetic spin labeled samples were however broader that those of the label-free samples. Despite overall structure conservation, this suggests that the spin labels induced some structural disorder.

The ^{1}H-^{15}N HSQC spectra measured for the $P_{127-205}$ S143C and S156C mutants with a paramagnetic spin label displayed severe line broadening (Figure 6B) due the contribution of the paramagnetic center to nuclear relaxation of nuclei within a 15 Å distance [65]. We measured the paramagnetic relaxation enhancement (PRE) for each residue, using the intensity ratio I_{para}/I_{dia} of NMR signals measured from paramagnetic and diamagnetic samples (Figure 6C). As expected, signals disappeared for residues up to 15 residues N- and C-terminally to the paramagnetic label position. For S143C, additional severe line broadening was observed in the 165–200 region. This points at transient long range interactions between the $P_{C\alpha}$ helices and the N-terminal half of P_{OD}. For S156C, additional line broadening was observed in the 175–200 region, which is only possible if the $P_{C\alpha}$ region is highly flexible.

We calculated theoretical PREs with the Flexible-meccano software [50] from an ensemble of 1000 $P_{127-205}$ structures (Figure 6C). The structural models were generated by using the α-helical propensities determined above and a 7 Å contact distance between P_{OD} residues 143–156 and $P_{C\alpha}$ residues 160–199 (Figure 6C). The calculated PREs approximated rather well the trend observed in experimental data. The dimensions measured for these $P_{127-205}$ structures are ~70 Å, which is compatible with a radius of gyration of 36.8 Å determined by SAXS. The structural ensemble of the $P_{C\alpha}$ region is diverse with helices of various lengths at different positions. These transient and flexible helices explore a large conformational space with respect to the P_{OD} domain and make transient contacts with P_{OD}, which lead to overall compaction of $P_{127-205}$.

3.6. Effect of Small Molecule Binding on the Dynamic Behavior of RSV P

Garcinol is a polycyclic polyprenylated acylphloroglucinol (PPAP) (Figure 7A). This natural product was isolated from *Moronobea coccinea* (Clusiaceae), a plant of French Guiana [66]. The plant is used in traditional medicine for its antioxidant and potentially antitumor activities [67]. Garcinol was reported to display activity against infectious agents such as Influenza A virus [68]. We identified garcinol as a potential RSV N-P interaction inhibitor by in vitro screening of an in-house chemical library. Garcinol exhibited cytotoxicity, but the in vitro results raised the question if it could directly interact with RSV N or P proteins.

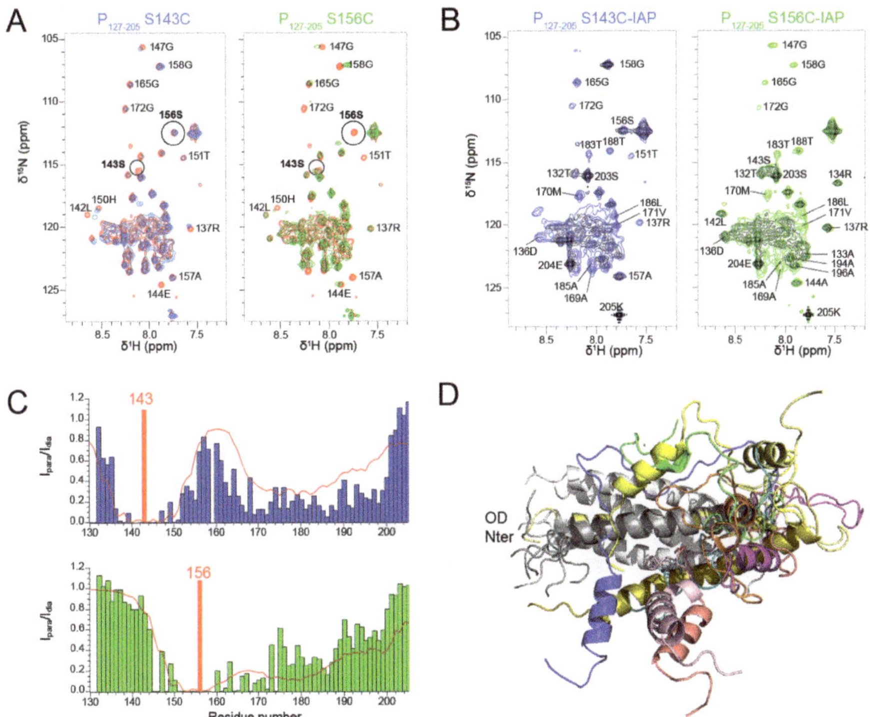

Figure 6. Spatial sampling of the C-terminal $P_{C\alpha}$ region proximal to the oligomerization domain of RSV P by Paramagnetic Relaxation Enhancement experiments. (**A**) Comparison between ^{1}H-^{15}N HSQC spectra of WT $P_{127-205}$ (red) and $P_{127-205}$ mutants S143C (blue) and S156C (green). NMR spectra were acquired at 600 MHz ^{1}H frequency and a temperature of 313 K. Assignments of several isolated signals are indicated. (**B**) ^{1}H-^{15}N HSQC spectra of IAP-labeled $P_{127-205}$ S143C and S156C mutants in paramagnetic (black) and diamagnetic states (S143C in blue, S156C in green). (**C**) Paramagnetic relaxation enhancements (PRE), determined as intensity ratios I_{para}/I_{dia} for each residue, are plotted as bar diagrams. The position of the IAP spin labels is indicated by a red bar. Theoretical PREs, plotted as a red line, were calculated with the Flexible-meccano software [50] from an ensemble of 1000 $P_{127-205}$ structures, using a contact distance of 7 Å between P_{OD} residues 143–156 and $P_{C\alpha}$ residues 160–199. (**D**) 10 randomly chosen $P_{127-205}$ conformers were aligned onto the P_{OD} structure to exemplify the transient structure of $P_{C\alpha}$ and the transient contacts between the $P_{C\alpha}$ helices and P_{OD}.

To probe the interaction with RSV P, we first used the $P_{161-241}$ fragment that contains an N-binding site relevant for RSV *holo* polymerase complex formation [69]. We measured an ^{1}H-^{15}N HSQC spectrum after addition of 3 molar equivalents of garcinol in DMSO and compared it to the reference spectrum of $P_{161-241}$ with an equivalent volume of DMSO (Figure 7B). The temperature was set to 293 K, since we showed previously that this fragment consists of transient helices and disordered regions [43]. The signals of the fully disordered C-terminal tail remained unaffected (Figure 7B,C). In contrast, the whole set of signals corresponding to the $P_{C\alpha}$ domain (40 residues) displayed severe line broadening and reduced intensities (Figure 7B,C). This was surprising, since the size of the perturbed area is much larger than expected for the binding site of a 600 Da molecule.

We wondered if garcinol would have the same effect on a P_{OD}-containing fragment. To answer this question, we used the $P_{127-205}$ fragment that lacks the P_{Ctail}, which remained unaffected by garcinol in the $P_{161-241}$ construct. We worked under experimental conditions where the P_{OD} signals were detectable, i.e. at a temperature of 313 K. Garcinol also induced severe line broadening at a 1:1 molar ratio in the whole $P_{C\alpha}$ domain (Figure 7B,C), such as in the $P_{161-241}$ fragment.

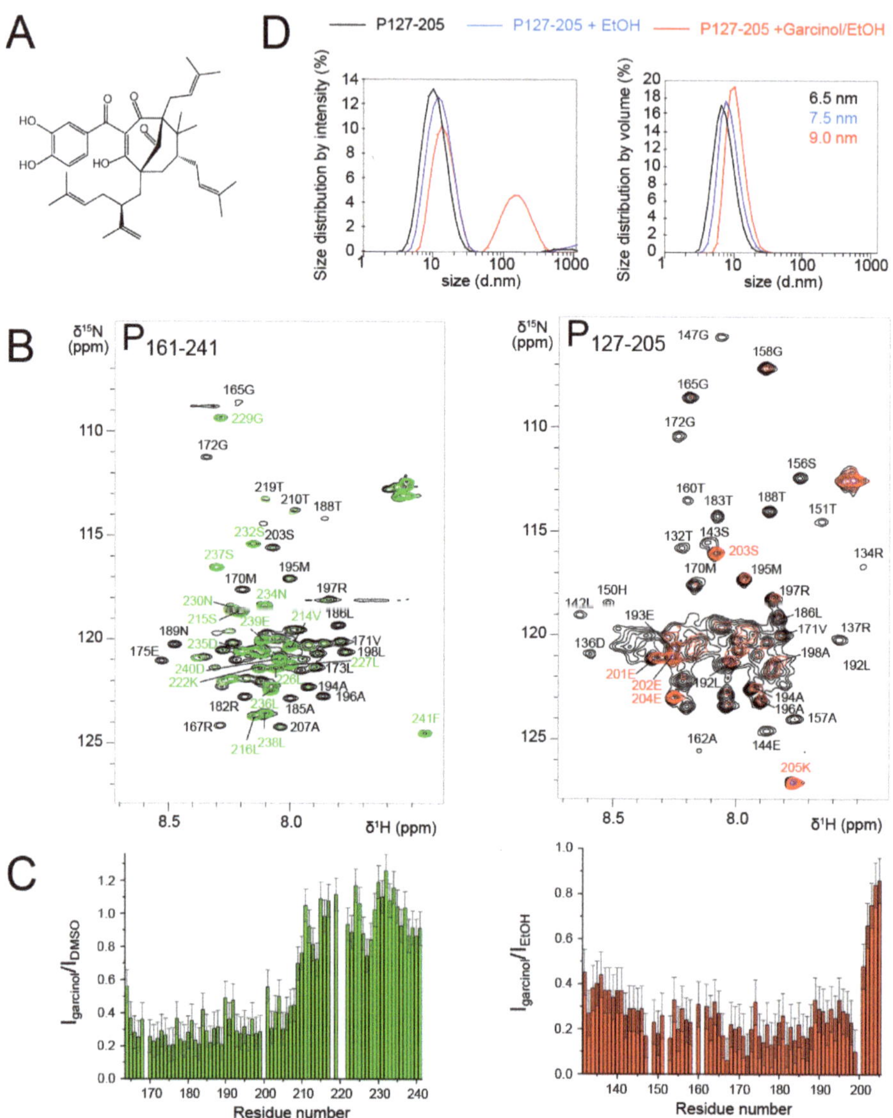

Figure 7. Effect of garcinol on the dynamics of RSV $P_{127–205}$ probed by NMR and DLS. (**A**) Chemical formula of garcinol. (**B**) Comparison between ^1H-^{15}N HSQC spectra of $P_{161–241}$ (T = 293 K, 150 μM) before (black) and after addition of 3 molar equivalents of garcinol in DMSO (green) and of $P_{127–205}$ (T = 313 K, 300 μM) before (black) and after 1 molar equivalent of garcinol in ethanol (red). (**C**) Intensity ratios, represented as bar diagrams, were determined from the intensities of ^1H-^{15}N HSQC spectra of $P_{127–205}$ and $P_{161–241}$ in the presence of garcinol versus solvent. Error bars correspond to the values of the RMSD determined in the regions with line broadening. (**D**) Size distribution in $P_{127–205}$ samples by dynamic light scattering.

Line broadening may arise from conformational exchange on the μs–ms timescale. In the case of $P_{C\alpha}$, which already contains transient helices, this would imply that garcinol triggers a change in local dynamics, either by affecting the exchange rate or by shifting the conformational equilibrium towards more order or disorder. Line broadening may also result from a change in molecular tumbling associated with formation of bigger sized particles. Garcinol induced line broadening in $P_{C\alpha}$ reminds of that observed for the P_{OD}

region, when it is flanked by large N- or C-terminal regions. This would imply that garcinol stabilizes the $P_{C\alpha}$ helices and/or promotes interprotomer or intermolecular interactions between these helices, leading to the observed effect. Our experiments did not permit the determination of the binding site of garcinol nor the stoichiometry of the putative garcinol-$P_{C\alpha}$ complex.

To acquire insight into potential self-association, we performed DLS measurements with $P_{127-205}$ after NMR measurements. From the size distribution by volume analysis, we deduced a hydrodynamic radius of ~32.5 Å (6.5 nm size, Figure 7D), which is in accordance with the 36.8 Å radius of gyration determined by SAXS. In the presence of ethanol and garcinol/ethanol, the apparent size (equivalent to a diameter) increased from 65 Å to 75 Å and 90 Å, respectively (Figure 7D). In the presence of garcinol/ethanol, a ~100 nm sized species was also observed in the distribution by intensity. However, according to the distribution by volume, it remains a minor species. We expect that an inter-protomer association would lead to compaction rather than to expansion of the $P_{127-205}$ structure. Destabilization of the helical structure could increase the hydrodynamic radius. However, an order-to-disorder transition would not lead to the nearly complete line broadening observed for $P_{161-241}$. The DLS data thus rather suggest that garcinol induces formation of higher order oligomeric forms of $P_{127-205}$, which are in equilibrium with the tetrameric form. We made a last control by adding the RSV N protein in the form of N-RNA rings to the $P_{161-241}$ fragment in the presence of garcinol. N bound to P exactly in the same manner as without garcinol, by involving the last C-terminal residues of P [43].

Taken together these results indicate that the dynamic state and the apparent size of $P_{127-205}$ are modulated by the presence of a molecule such as garcinol. From our data we cannot conclude on the exact molecular mechanism. However, changes are mediated by the isolated transient helices in the $P_{C\alpha}$ domain. Given the importance of $P_{C\alpha}$ for the formation of the RSV P-L complex, such molecules may be used to interfere with the viral polymerase by precluding the P-L interaction.

4. Discussion and Conclusions

The RSV phosphoprotein acts as a hub protein for the viral RNA dependent RNA polymerase complex by assembling its different components [17]. It binds to the catalytic subunit L on the one hand and to the nucleoprotein enchasing the viral RNA genome on the other hand to form the holo polymerase complex [70,71]. It recruits additional factors such as the RSV M2-1 transcription antiterminator to the polymerase complex [72]. It serves as a docking platform for tertiary complexes such as the complex formed with the protein phosphatase 1 and M2-1, leading to dephosphorylation of M2-1 [44]. The phosphoprotein also plays a role in viral budding. We have recently shown that P directly interacts with the RSV matrix protein and that the tetrameric nature of RSV P is a requirement for this function [73].

The RSV P protein has long eluded structural elucidation due to its intrinsic disorder [36,39,74]. We and others investigated its overall structural and dynamic properties in solution as well as disorder-order transitions of its different domains using several biophysical approaches that reported either on domain properties or individual amino acids in the case of NMR [38,43]. Recently, high-resolution structural data was obtained by cryo-EM for the RSV polymerase, including RSV P, as the main co-factor of the large catalytic subunit L [41,42]. The L—P complex structure highlighted two structural aspects about the P protein. First, its N-terminal domain was invisible in the cryo-EM maps, indicating that it remained disordered and unbound. Second, the domain with high α-helical propensity, $P_{C\alpha}$, became ordered upon binding L, but each protomer adopted a specific binding mode, with different structures, i.e., containing a different number of α-helices with different boundaries and different binding sites (Figure 1B) [17,41].

Here we investigated into more details the structure and dynamics of the consecutive P_{OD} and $P_{C\alpha}$ domains in unbound RSV P. $P_{C\alpha}$ is a neither fully disordered nor globular region, but situated in the continuum between these two extremes (Figures 3 and 4). P_{OD}

is the only stably folded part of the protein. However, even the boundaries of P_{OD} are fluctuating. Structure resolution by solution NMR showed that the C-terminal end of the coiled coil tetramer may extend up to residue V154 (Figure 5), whereas in the L—P complex, one protomer already disengages from the tetramer one helix turn before. We showed that transient α-helices were formed in unbound P protein, by using tetrameric fragments of P. In previous studies we had shown that these helices also exist in monomeric fragments [43]. We moreover showed that the full $P_{C\alpha}$ domain was necessary to promote α-helicity, since $P_{C\alpha}$ did not adopt its native conformation in the truncated $P_{127\text{-}184}$ fragment (Figure 3). The boundaries of these transient helices are delineating those of the stabilized helices in the L—P complex. Interestingly the α_{C1} region displayed the highest α-helical propensities (Figure 3). α_{C1} was also singled out in high-pressure NMR experiments by higher linear and quadratic pressure coefficients (Figure 4). The α_{C1} region is the only one adopting α-helical secondary structure in all four protomers in the RSV L—P complex [41,42]. From our data, we conclude that the bound conformation of α_{C1} pre-exists in unbound P, in equilibrium with a more disordered state shifted towards the ordered state.

By using paramagnetic labels, we showed that the $P_{C\alpha}$ domain makes transient contacts with P_{OD}, leading to compaction of the $P_{127\text{-}205}$ fragment. The overall size of the structural models derived from PRE analysis (Figure 6) agrees with the radius of gyration determined by SAXS and the size measured by DLS. Since we previously showed that the last 30–35 C-terminal residues of the P protein, corresponding to P_{Ctail}, are fully disordered [43], we may infer that these transient contacts and overall compaction, which are inherent to the tetrameric nature of RSV P, also take place in the full-length protein.

Finally, we showed that the $P_{C\alpha}$ domain might be targeted by small molecules to modify its dynamics Figure 7), by promoting self-association of the free helices from different protomers or by promoting oligomerization of P. The structural plasticity of the P protein, and in particular of its $P_{C\alpha}$ domain, plays a crucial role for the L—P complex formation, i.e., for a functional viral polymerase. This now needs to be verified for the full-length P protein using molecules with demonstrated antiviral activity.

Supplementary Materials: The following are available online at https://www.mdpi.com/article/10.3390/biom11081225/s1, Table S1: NMR restraints and structural statistics for RSV $P_{127\text{-}163}$.

Author Contributions: Conceptualization, methodology, C.S.; formal analysis, C.C., C.-M.C. and C.S.; investigation, A.T., C.C., C.-M.C., C.S., F.B., M.B., M.G. and M.L.; data curation, A.T., B.B., C.C. and C.S.; writing—original draft preparation, C.S.; writing—review and editing, A.T., B.B., C.C., C.-M.C., C.S., F.B., J.-F.E., M.B., M.G. and M.L.; visualization, C.C. and C.S.; supervision, C.S.; funding acquisition, C.S. and J.-F.E. All authors have read and agreed to the published version of the manuscript.

Funding: This work was supported by Agence Nationale de la Recherche (grants ANR-13-ISV3-0007, ANR-19-CE18-0012-01) and LabEx LERMIT (C.S.), Région Ile-de-France DIM Malinf (doctoral fellowship to C.C.) and Université Paris Saclay (doctoral fellowship to C.M.C.) and the INCEPTION project (PIA/ANR-16-CONV-0005). SAXS experiments were performed on the SWING beamline at SOLEIL Synchroton, France (proposal 20181072). Financial support from the IR-RMN-THC FR 3050 CNRS for conducting the research is gratefully acknowledged.

Institutional Review Board Statement: Not applicable.

Informed Consent Statement: Not applicable.

Data Availability Statement: Data is contained within the article.

Acknowledgments: We thank Annie Moretto for valuable technical assistance in the wetlab and François Giraud for maintenance of the NMR equipment.

Conflicts of Interest: The authors declare no conflict of interest. The funders had no role in the design of the study; in the collection, analyses or interpretation of data; in the writing of the manuscript or in the decision to publish the results.

References

1. Brocca, S.; Grandori, R.; Longhi, S.; Uversky, V. Liquid–liquid phase separation by intrinsically disordered protein regions of viruses: Roles in viral life cycle and control of virus–host interactions. *Int. J. Mol. Sci.* **2020**, *21*, 9045. [CrossRef]
2. Mishra, P.M.; Verma, N.C.; Rao, C.; Uversky, V.N.; Nandi, C.K. Intrinsically disordered proteins of viruses: Involvement in the mechanism of cell regulation and pathogenesis. In *Dancing Protein Clouds: Intrinsically Disordered Proteins in Health and Disease, Part B*; Elsevier: Amsterdam, The Netherlands, 2020; pp. 1–78. [CrossRef]
3. Kumar, N.; Kaushik, R.; Tennakoon, C.; Uversky, V.N.; Longhi, S.; Zhang, K.Y.J.; Bhatia, S. Comprehensive intrinsic disorder analysis of 6108 Viral proteomes: From the extent of intrinsic disorder penetrance to functional annotation of disordered viral proteins. *J. Proteome Res.* **2021**, *20*, 2704–2713. [CrossRef]
4. Wright, P.E.; Dyson, H.J. Linking folding and binding. *Curr. Opin. Struct. Biol.* **2009**, *19*, 31–38. [CrossRef]
5. Uversky, V.N. Unusual biophysics of intrinsically disordered proteins. *Biochim. Biophys. Acta* **2013**, *1834*, 932–951. [CrossRef]
6. Sharma, R.; Raduly, Z.; Miskei, M.; Fuxreiter, M. Fuzzy complexes: Specific binding without complete folding. *FEBS Lett.* **2015**, *589*, 2533–2542. [CrossRef]
7. Tompa, P.; Schad, E.; Tantos, A.; Kalmar, L. Intrinsically disordered proteins: Emerging interaction specialists. *Curr. Opin. Struct. Biol.* **2015**, *35*, 49–59. [CrossRef] [PubMed]
8. Tompa, P.; Fuxreiter, M. Fuzzy complexes: Polymorphism and structural disorder in protein-protein interactions. *Trends Biochem. Sci.* **2008**, *33*, 2–8. [CrossRef]
9. Ogino, T.; Green, T.J. RNA synthesis and capping by non-segmented negative strand RNA viral polymerases: Lessons from a prototypic virus. *Front. Microbiol.* **2019**, *10*, 1490. [CrossRef] [PubMed]
10. Fearns, R. The respiratory syncytial virus polymerase: A multitasking machine. *Trends Microbiol.* **2019**, *27*, 969–971. [CrossRef]
11. Fearns, R.; Plemper, R.K. Polymerases of paramyxoviruses and pneumoviruses. *Virus Res.* **2017**, *234*, 87–102. [CrossRef] [PubMed]
12. Morin, B.; Kranzusch, P.J.; Rahmeh, A.A.; Whelan, S.P. The polymerase of negative-stranded RNA viruses. *Curr. Opin. Virol.* **2013**, *3*, 103–110. [CrossRef]
13. Ruigrok, R.W.; Crepin, T.; Kolakofsky, D. Nucleoproteins and nucleocapsids of negative-strand RNA viruses. *Curr. Opin. Microbiol.* **2011**, *14*, 504–510. [CrossRef] [PubMed]
14. Galloux, M.; Tarus, B.; Blazevic, I.; Fix, J.; Duquerroy, S.; Eleouet, J.F. Characterization of a viral phosphoprotein binding site on the surface of the respiratory syncytial nucleoprotein. *J. Virol.* **2012**, *86*, 8375–8387. [CrossRef] [PubMed]
15. Habchi, J.; Blangy, S.; Mamelli, L.; Jensen, M.R.; Blackledge, M.; Darbon, H.; Oglesbee, M.; Shu, Y.; Longhi, S. Characterization of the interactions between the nucleoprotein and the phosphoprotein of henipavirus. *J. Biol. Chem.* **2011**, *286*, 13583–13602. [CrossRef] [PubMed]
16. Shu, Y.; Habchi, J.; Costanzo, S.; Padilla, A.; Brunel, J.; Gerlier, D.; Oglesbee, M.; Longhi, S. Plasticity in structural and functional interactions between the phosphoprotein and nucleoprotein of measles virus. *J. Biol. Chem.* **2012**, *287*, 11951–11967. [CrossRef]
17. Cardone, C.; Caseau, C.-M.; Pereira, N.; Sizun, C. Pneumoviral phosphoprotein, a multidomain adaptor-like protein of apparent low structural complexity and high conformational versatility. *Int. J. Mol. Sci.* **2021**, *22*, 1537. [CrossRef]
18. Karlin, D.; Belshaw, R. Detecting remote sequence homology in disordered proteins: Discovery of conserved motifs in the n-termini of mononegavirales phosphoproteins. *PLoS ONE* **2012**, *7*, e31719. [CrossRef] [PubMed]
19. Longhi, S. Structural disorder within paramyxoviral nucleoproteins. *FEBS Lett.* **2015**, *589*, 2649–2659. [CrossRef]
20. Habchi, J.; Longhi, S. Structural disorder within paramyxoviral nucleoproteins and phosphoproteins in their free and bound forms: From predictions to experimental assessment. *Int. J. Mol. Sci.* **2015**, *16*, 15688–15726. [CrossRef]
21. Guseva, S.; Milles, S.; Blackledge, M.; Ruigrok, R.W.H. The nucleoprotein and phosphoprotein of measles virus. *Front. Microbiol.* **2019**, *10*, 1832. [CrossRef]
22. Jensen, M.R.; Communie, G.; Ribeiro, E.A., Jr.; Martinez, N.; Desfosses, A.; Salmon, L.; Mollica, L.; Gabel, F.; Jamin, M.; Longhi, S.; et al. Intrinsic disorder in measles virus nucleocapsids. *Proc. Natl. Acad. Sci. USA* **2011**, *108*, 9839–9844. [CrossRef] [PubMed]
23. Communie, G.; Habchi, J.; Yabukarski, F.; Blocquel, D.; Schneider, R.; Tarbouriech, N.; Papageorgiou, N.; Ruigrok, R.W.; Jamin, M.; Jensen, M.R.; et al. Atomic resolution description of the interaction between the nucleoprotein and phosphoprotein of Hendra virus. *PLoS Pathog.* **2013**, *9*, e1003631. [CrossRef]
24. Schneider, R.; Maurin, D.; Communie, G.; Kragelj, J.; Hansen, D.F.; Ruigrok, R.W.; Jensen, M.R.; Blackledge, M. Visualizing the molecular recognition trajectory of an intrinsically disordered protein using multinuclear relaxation dispersion NMR. *J. Am. Chem. Soc.* **2015**, *137*, 1220–1229. [CrossRef] [PubMed]
25. Nikolic, J.; Le Bars, R.; Lama, Z.; Scrima, N.; Lagaudrière-Gesbert, C.; Gaudin, Y.; Blondel, D. Negri bodies are viral factories with properties of liquid organelles. *Nat. Commun.* **2017**, *8*, 1–13. [CrossRef]
26. Heinrich, B.S.; Maliga, Z.; Stein, D.A.; Hyman, A.A.; Whelan, S.P.J.; Palese, P. Phase transitions drive the formation of vesicular stomatitis virus replication compartments. *mBIO* **2018**, *9*, e02290-17. [CrossRef] [PubMed]
27. Su, J.; Wilson, M.; Samuel, C.; Ma, D. Formation and function of liquid-like viral factories in negative-sense single-stranded RNA virus infections. *Viruses* **2021**, *13*, 126. [CrossRef]
28. Guseva, S.; Milles, S.; Ringkobing Jensen, M.; Salvi, N.; Kleman, J.-P.; Maurin, D.; Ruigrok, R.W.H.; Blackledge, M. Measles virus nucleo- and phosphoproteins form liquid-like phase-separated compartments that promote nucleocapsid assembly. *Sci. Adv.* **2020**, *6*, eaaz7095. [CrossRef]

29. Galloux, M.; Risso-Ballester, J.; Richard, C.A.; Fix, J.; Rameix-Welti, M.A.; Eleouet, J.F. Minimal elements required for the formation of respiratory syncytial virus cytoplasmic inclusion bodies in vivo and in vitro. *mBio* **2020**, *11*, e01202-20. [CrossRef]

30. Perch, S.G. Causes of severe pneumonia requiring hospital admission in children without HIV infection from Africa and Asia: The Perch multi-country case-control study. *Lancet* **2019**, *394*, 757–779. [CrossRef]

31. Coultas, J.A.; Smyth, R.; Openshaw, P.J. Respiratory syncytial virus (RSV): A scourge from infancy to old age. *Thorax* **2019**, *74*, 986–993. [CrossRef]

32. Collins, P.L.; Melero, J.A. Progress in understanding and controlling respiratory syncytial virus: Still crazy after all these years. *Virus Res.* **2011**, *162*, 80–99. [CrossRef]

33. Whelan, J.N.; Reddy, K.D.; Uversky, V.N.; Teng, M.N. Functional correlations of respiratory syncytial virus proteins to intrinsic disorder. *Mol. Biosyst.* **2016**, *12*, 1507–1526. [CrossRef] [PubMed]

34. Romero, P.; Obradovic, Z.; Li, X.; Garner, E.C.; Brown, C.J.; Dunker, A.K. Sequence complexity of disordered protein. *Proteins Struct. Funct. Genet.* **2001**, *42*, 38–48. [CrossRef]

35. Castagne, N.; Barbier, A.; Bernard, J.; Rezaei, H.; Huet, J.C.; Henry, C.; Da Costa, B.; Eleouet, J.F. Biochemical characterization of the respiratory syncytial virus P-P and P-N protein complexes and localization of the P protein oligomerization domain. *J. Gen. Virol.* **2004**, *85*, 1643–1653. [CrossRef]

36. Simabuco, F.M.; Asara, J.M.; Guerrero, M.C.; Libermann, T.A.; Zerbini, L.F.; Ventura, A.M. Structural analysis of human respiratory syncytial virus P protein: Identification of intrinsically disordered domains. *Braz. J. Microbiol.* **2011**, *42*, 340–345. [CrossRef] [PubMed]

37. Esperante, S.A.; Paris, G.; de Prat-Gay, G. Modular unfolding and dissociation of the human respiratory syncytial virus phosphoprotein p and its interaction with the m(2-1) antiterminator: A singular tetramer-tetramer interface arrangement. *Biochemistry* **2012**, *51*, 8100–8110. [CrossRef]

38. Noval, M.G.; Esperante, S.A.; Molina, I.G.; Chemes, L.B.; Prat-Gay, G. Intrinsic disorder to order transitions in the scaffold phosphoprotein P from the respiratory syncytial virus RNA-polymerase complex. *Biochemistry* **2016**, *55*, 1441–1454. [CrossRef] [PubMed]

39. Llorente, M.T.; Taylor, I.A.; Lopez-Vinas, E.; Gomez-Puertas, P.; Calder, L.J.; Garcia-Barreno, B.; Melero, J.A. Structural properties of the human respiratory syncytial virus P protein: Evidence for an elongated homotetrameric molecule that is the smallest orthologue within the family of paramyxovirus polymerase cofactors. *Proteins* **2008**, *72*, 946–958. [CrossRef] [PubMed]

40. Asenjo, A.; Mendieta, J.; Gomez-Puertas, P.; Villanueva, N. Residues in human respiratory syncytial virus P protein that are essential for its activity on RNA viral synthesis. *Virus Res.* **2008**, *132*, 160–173. [CrossRef]

41. Gilman, M.S.A.; Liu, C.; Fung, A.; Behera, I.; Jordan, P.; Rigaux, P.; Ysebaert, N.; Tcherniuk, S.; Sourimant, J.; Eleouet, J.F.; et al. Structure of the respiratory syncytial virus polymerase complex. *Cell* **2019**, *179*, 193–204. [CrossRef] [PubMed]

42. Cao, D.; Gao, Y.; Roesler, C.; Rice, S.; D'Cunha, P.; Zhuang, L.; Slack, J.; Domke, M.; Antonova, A.; Romanelli, S.; et al. Cryo-EM structure of the respiratory syncytial virus RNA polymerase. *Nat. Commun.* **2020**, *11*, 368. [CrossRef] [PubMed]

43. Pereira, N.; Cardone, C.; Lassoued, S.; Galloux, M.; Fix, J.; Assrir, N.; Lescop, E.; Bontems, F.; Eleouet, J.F.; Sizun, C. New insights into structural disorder in human respiratory syncytial virus phosphoprotein and implications for binding of protein partners. *J. Biol. Chem.* **2017**, *292*, 2120–2131. [CrossRef]

44. Richard, C.A.; Rincheval, V.; Lassoued, S.; Fix, J.; Cardone, C.; Esneau, C.; Nekhai, S.; Galloux, M.; Rameix-Welti, M.A.; Sizun, C.; et al. RSV hijacks cellular protein phosphatase 1 to regulate M2-1 phosphorylation and viral transcription. *PLoS Pathog.* **2018**, *14*, e1006920. [CrossRef] [PubMed]

45. Vranken, W.F.; Boucher, W.; Stevens, T.J.; Fogh, R.H.; Pajon, A.; Llinas, M.; Ulrich, E.L.; Markley, J.L.; Ionides, J.; Laue, E.D. The CCPN data model for NMR spectroscopy: Development of a software pipeline. *Proteins* **2005**, *59*, 687–696. [CrossRef]

46. Favier, A.; Brutscher, B. NMRlib: User-friendly pulse sequence tools for Bruker NMR spectrometers. *J. Biomol. NMR* **2019**, *73*, 199–211. [CrossRef]

47. Shen, Y.; Bax, A. Protein backbone and sidechain torsion angles predicted from NMR chemical shifts using artificial neural networks. *J. Biomol. NMR* **2013**, *56*, 227–241. [CrossRef] [PubMed]

48. Bardiaux, B.; Malliavin, T.; Nilges, M. ARIA for solution and solid-state NMR. *Methods Mol. Biol.* **2012**, *831*, 453–483. [CrossRef] [PubMed]

49. Nilges, M. A calculation strategy for the structure determination of symmetric demers by1H NMR. *Proteins Struct. Funct. Genet.* **1993**, *17*, 297–309. [CrossRef]

50. Ozenne, V.; Bauer, F.; Salmon, L.; Huang, J.R.; Jensen, M.R.; Segard, S.; Bernado, P.; Charavay, C.; Blackledge, M. Flexible-meccano: A tool for the generation of explicit ensemble descriptions of intrinsically disordered proteins and their associated experimental observables. *Bioinformatics* **2012**, *28*, 1463–1470. [CrossRef]

51. Perez, J.; Nishino, Y. Advances in X-ray scattering: From solution SAXS to achievements with coherent beams. *Curr. Opin. Struct. Biol.* **2012**, *22*, 670–678. [CrossRef]

52. David, G.; Perez, J. Combined sampler robot and high-performance liquid chromatography: A fully automated system for biological small-angle X-ray scattering experiments at the Synchrotron SOLEIL SWING beamline. *J. Appl. Cryst.* **2009**, *42*, 892–900. [CrossRef]

53. Brookes, E.; Vachette, P.; Rocco, M.; Perez, J. US-SOMO HPLC-SAXS module: Dealing with capillary fouling and extraction of pure component patterns from poorly resolved SEC-SAXS data. *J. Appl. Crystallogr.* **2016**, *49*, 1827–1841. [CrossRef]

54. Konarev, P.V.; Volkov, V.V.; Sokolova, A.V.; Koch, M.H.; Svergun, D.I. PRIMUS: A Windows PC-based system for small-angle scattering data analysis. *J. Appl. Cryst.* **2003**, *36*, 1277–1282. [CrossRef]

55. Schneidman-Duhovny, D.; Hammel, M.; Sali, A. FoXS: A web server for rapid computation and fitting of SAXS profiles. *Nucleic Acids Res.* **2010**, *38*, W540–W544. [CrossRef] [PubMed]

56. Schrödinger, LCC. *The PyMOL Molecular Graphics System*; Version 2.4; Schrödinger, LCC: New York, NY, USA, 2010.

57. Roche, J.; Royer, C.A.; Roumestand, C. Monitoring protein folding through high pressure NMR spectroscopy. *Prog. Nucl. Magn. Reson. Spectrosc.* **2017**, *102–103*, 15–31. [CrossRef]

58. Kitahara, R.; Hata, K.; Li, H.; Williamson, M.P.; Akasaka, K. Pressure-induced chemical shifts as probes for conformational fluctuations in proteins. *Prog. Nucl. Magn. Reson. Spectrosc.* **2013**, *71*, 35–58. [CrossRef]

59. Caro, J.A.; Wand, A.J. Practical aspects of high-pressure NMR spectroscopy and its applications in protein biophysics and structural biology. *Methods* **2018**, *148*, 67–80. [CrossRef] [PubMed]

60. Koehler, J.; Beck Erlach, M.; Crusca, E.; Kremer, W.; Munte, C.E.; Kalbitzer, H.R. Pressure dependence of 15N chemical shifts in model peptides Ac-Gly-Gly-X-Ala-NH2. *Materials* **2012**, *5*, 1774–1786. [CrossRef]

61. Quinlan, R.J.; Reinhart, G.D. Baroresistant buffer mixtures for biochemical analyses. *Anal. Biochem.* **2005**, *341*, 69–76. [CrossRef]

62. Rocchia, W.; Alexov, E.; Honig, B. Extending the applicability of the nonlinear Poisson-Boltzmann equation: Multiple dielectric constants and multivalent ions. *J. Phys. Chem. B* **2001**, *105*, 6507–6514. [CrossRef]

63. Receveur-Brechot, V.; Durand, D. How random are intrinsically disordered proteins? A small angle scattering perspective. *Curr. Protein Pept. Sci.* **2012**, *13*, 55–75. [CrossRef] [PubMed]

64. Clore, G.M.; Iwahara, J. Theory, practice, and applications of paramagnetic relaxation enhancement for the characterization of transient low-population states of biological macromolecules and their complexes. *Chem. Rev.* **2009**, *109*, 4108–4139. [CrossRef]

65. Lietzow, M.A.; Jamin, M.; Dyson, H.J.; Wright, P.E. Mapping long-range contacts in a highly unfolded protein. *J. Mol. Biol.* **2002**, *322*, 655–662. [CrossRef]

66. Marti, G.; Eparvier, V.; Moretti, C.; Susplugas, S.; Prado, S.; Grellier, P.; Retailleau, P.; Guéritte, F.; Litaudon, M. Antiplasmodial benzophenones from the trunk latex of Moronobea coccinea (Clusiaceae). *Phytochemistry* **2009**, *70*, 75–85. [CrossRef]

67. Padhye, S.; Ahmad, A.; Oswal, N.; Sarkar, F.H. Emerging role of Garcinol, the antioxidant chalcone from Garcinia indica Choisy and its synthetic analogs. *J. Hematol. Oncol.* **2009**, *2*, 1–13. [CrossRef] [PubMed]

68. Schobert, R.; Biersack, B. Chemical and biological aspects of garcinol and isogarcinol: Recent developments. *Chem. Biodivers.* **2019**, *16*, e1900366. [CrossRef]

69. Ouizougun-Oubari, M.; Pereira, N.; Tarus, B.; Galloux, M.; Lassoued, S.; Fix, J.; Tortorici, M.A.; Hoos, S.; Baron, B.; England, P.; et al. A druggable pocket at the nucleocapsid/phosphoprotein interaction site of human respiratory syncytial virus. *J. Virol.* **2015**, *89*, 11129–11143. [CrossRef] [PubMed]

70. Galloux, M.; Gabiane, G.; Sourimant, J.; Richard, C.A.; England, P.; Moudjou, M.; Aumont-Nicaise, M.; Fix, J.; Rameix-Welti, M.A.; Eleouet, J.F. Identification and characterization of the binding site of the respiratory syncytial virus phosphoprotein to RNA-free nucleoprotein. *J. Virol.* **2015**, *89*, 3484–3496. [CrossRef] [PubMed]

71. Sourimant, J.; Rameix-Welti, M.A.; Gaillard, A.L.; Chevret, D.; Galloux, M.; Gault, E.; Eleouet, J.F. Fine mapping and characterization of the L-polymerase-binding domain of the respiratory syncytial virus phosphoprotein. *J. Virol.* **2015**, *89*, 4421–4433. [CrossRef]

72. Blondot, M.L.; Dubosclard, V.; Fix, J.; Lassoued, S.; Aumont-Nicaise, M.; Bontems, F.; Eleouet, J.F.; Sizun, C. Structure and functional analysis of the RNA and viral phosphoprotein binding domain of respiratory syncytial virus M2-1 protein. *PLoS Pathog.* **2012**, *8*, e1002734. [CrossRef] [PubMed]

73. Bajorek, M.; Galloux, M.; Richard, C.-A.; Szekely, O.; Rosenzweig, R.; Sizun, C.; Eleouet, J.-F. Tetramerization of phosphoprotein is essential for respiratory syncytial virus budding while its n terminal region mediates direct interactions with the matrix protein. *J. Virol.* **2021**, *95*, e02217-20. [CrossRef] [PubMed]

74. Llorente, M.T.; Garcia-Barreno, B.; Calero, M.; Camafeita, E.; Lopez, J.A.; Longhi, S.; Ferron, F.; Varela, P.F.; Melero, J.A. Structural analysis of the human respiratory syncytial virus phosphoprotein: Characterization of an alpha-helical domain involved in oligomerization. *J. Gen. Virol.* **2006**, *87*, 159–169. [CrossRef] [PubMed]

 *biomolecules*

Article

Stabilization Effect of Intrinsically Disordered Regions on Multidomain Proteins: The Case of the Methyl-CpG Protein 2, MeCP2

David Ortega-Alarcon [1], Rafael Claveria-Gimeno [1,2,3,†], Sonia Vega [1], Olga C. Jorge-Torres [4], Manel Esteller [4,5,6,7], Olga Abian [1,2,3,8,9,*] and Adrian Velazquez-Campoy [1,3,8,9,10,*]

[1] Institute of Biocomputation and Physics of Complex Systems (BIFI), Joint Units IQFR-CSIC-BIFI, and GBsC-CSIC-BIFI, Universidad de Zaragoza, 50018 Zaragoza, Spain; dortega@bifi.es (D.O.-A.); rafacg@certest.es (R.C.-G.); svega@bifi.es (S.V.)

[2] Instituto Aragonés de Ciencias de la Salud (IACS), 50009 Zaragoza, Spain

[3] Instituto de Investigación Sanitaria de Aragón (IIS Aragón), 50009 Zaragoza, Spain

[4] Josep Carreras Leukaemia Research Institute (IJC), 08916 Badalona, Spain; ojorge@carrerasresearch.org (O.C.J.-T.); mesteller@carrerasresearch.org (M.E.)

[5] Centro de Investigacion Biomedica en Red Cancer (CIBERONC), 28029 Madrid, Spain

[6] Institucio Catalana de Recerca i Estudis Avançats (ICREA), 08010 Barcelona, Spain

[7] Physiological Sciences Department, School of Medicine and Health Sciences, University of Barcelona (UB), l'Hospitalet de Llobregat, 08907 Barcelona, Spain

[8] Centro de Investigación Biomédica en Red en el Área Temática de Enfermedades Hepáticas y Digestivas (CIBERehd), 28029 Madrid, Spain

[9] Departamento de Bioquímica y Biología Molecular y Celular, Universidad de Zaragoza, 50009 Zaragoza, Spain

[10] Fundación ARAID, Gobierno de Aragón, 50009 Zaragoza, Spain

[*] Correspondence: oabifra@unizar.es (O.A.); adrianvc@unizar.es (A.V.-C.); Tel.: +34-876-555417 (O.A.); +34-976-762996 (A.V.-C.)

[†] Current address: Certest Biotec S.L., 50840 Zaragoza, Spain.

Citation: Ortega-Alarcon, D.; Claveria-Gimeno, R.; Vega, S.; Jorge-Torres, O.C.; Esteller, M.; Abian, O.; Velazquez-Campoy, A. Stabilization Effect of Intrinsically Disordered Regions on Multidomain Proteins: The Case of the Methyl-CpG Protein 2, MeCP2. *Biomolecules* **2021**, *11*, 1216. https://doi.org/10.3390/biom11081216

Academic Editors: Simona Maria Monti, Giuseppina De Simone and Emma Langella

Received: 10 July 2021
Accepted: 13 August 2021
Published: 16 August 2021

Publisher's Note: MDPI stays neutral with regard to jurisdictional claims in published maps and institutional affiliations.

Abstract: Intrinsic disorder plays an important functional role in proteins. Disordered regions are linked to posttranslational modifications, conformational switching, extra/intracellular trafficking, and allosteric control, among other phenomena. Disorder provides proteins with enhanced plasticity, resulting in a dynamic protein conformational/functional landscape, with well-structured and disordered regions displaying reciprocal, interdependent features. Although lacking well-defined conformation, disordered regions may affect the intrinsic stability and functional properties of ordered regions. MeCP2, methyl-CpG binding protein 2, is a multifunctional transcriptional regulator associated with neuronal development and maturation. MeCP2 multidomain structure makes it a prototype for multidomain, multifunctional, intrinsically disordered proteins (IDP). The methyl-binding domain (MBD) is one of the key domains in MeCP2, responsible for DNA recognition. It has been reported previously that the two disordered domains flanking MBD, the N-terminal domain (NTD) and the intervening domain (ID), increase the intrinsic stability of MBD against thermal denaturation. In order to prove unequivocally this stabilization effect, ruling out any artifactual result from monitoring the unfolding MBD with a local fluorescence probe (the single tryptophan in MBD) or from driving the protein unfolding by temperature, we have studied the MBD stability by differential scanning calorimetry (reporting on the global unfolding process) and chemical denaturation (altering intramolecular interactions by a different mechanism compared to thermal denaturation).

Keywords: MeCP2; intrinsically disordered proteins; structural stability; thermal and chemical denaturation; differential scanning calorimetry

Biomolecules **2021**, *11*, 1216. https://doi.org/10.3390/biom11081216

https://www.mdpi.com/journal/biomolecules

1. Introduction

Intrinsically disordered proteins (IDP) and intrinsically disordered regions (IDR) are involved in many physiological mechanisms and pathologies. Although much effort has been dedicated towards understanding their structural and functional properties, IDPs and IDRs remain largely elusive. IDPs and IDRs contain a large proportion of polar, hydrophilic residues, compared to well-folded proteins and regions, which provide them the capability of populating a dynamic, rich ensemble of structures [1,2]. They do not fold spontaneously into a stable conformation in aqueous solvent because they lack a hydrophobic core or it is insufficiently large, but they are usually very susceptible to environmental factors (e.g., presence of ligands and interacting macromolecules, redox state, or pH of the surroundings) and posttranslational modifications [3]. Interestingly, they may fold into well-defined conformations when interacting with a biological partner or under appropriate conditions (the binding partner or concomitant processes, such as electron transfer or de/protonation, provide the additional physico-chemical context for establishing stabilizing interactions and triggering the disorder-to-order transition) or remain disordered ("fuzzy" complexes).

Disorder in proteins provides some unexpected advantages. First, disorder increases the structural plasticity and flexibility, allowing proteins to adopt multiple conformations within a complex conformational landscape depending on the environmental conditions, which is important for interacting with many biological partners [4,5] (in fact, IDPs or proteins containing IDRs represent important hubs in metabolic and signaling networks) and undergoing conformational changes required for cellular trafficking, internalization, and degradation [6,7]. Second, disorder makes some regions more susceptible to posttranslational modifications [8], which represents another protein regulation level conditioning the accessible conformations and the potential interactions with other biomolecules. Thirdly, disorder endows proteins with an additional regulation level based on allosteric control [9–12], which broadly consists in the regulation of the conformational landscape by ligand interaction (where ligand is any molecule interacting with a given protein: ion, small molecule, macromolecule.) [13,14].

Protein disorder has been maintained and exploited by evolution [15–18]. Therefore, it must be a key property that must be understood and accounted for when investigating structural and functional features in proteins. Some consequences derived from the presence of IDRs and the direct connection between order/disorder and function in proteins are not fully understood yet. Disordered regions in particular contribute to protein stability and function, sometimes with a counterbalancing effect [19,20], and should be considered as functional regions and not just flexible stretches [21,22].

The methyl-CpG binding protein 2, MeCP2, a multidomain protein containing a large proportion of intrinsic disorder and interacting with many biological partners, is a physiologically interesting and clinically relevant IDP. MeCP2 is involved in many key physiological processes associated with neuronal development, maturation, and plasticity [23,24], and MeCP2 deleterious mutations are associated with Rett syndrome, a neurodevelopmental disorder related to the autistic spectrum [25–27]. Although all domains must be relevant, because MeCP2 is a main transcriptional regulator and chromatin remodeling element, two domains emerge among the others: the methyl-binding domain (MBD) and the transcriptional repression domain (TRD). MBD is the domain responsible for the interaction with methylated-CpG-rich promoters [28–30]. Interestingly, MBD has a DNA binding site and contains one of the few well-structured regions in this protein, which is flanked by two completely disordered domains: N-terminal domain (NTD) and intervening domain (ID). Despite their lack of structure, both domains are critical for MeCP2 functions. In particular, mutations and posttranslational modifications in NTD affect and modulate MeCP2 functions [31,32]. In addition, the two identified MeCP2 isoforms differ in just a few residues located at the beginning of the NTD, but that small difference has important consequences in function, expression, and structural properties [33]. The ID not only provides a second DNA binding site, but it also considerably increases the DNA binding affinity of the MBD site.

It has been reported that both NTD and ID increase the intrinsic structural stability of MBD, as observed by thermal unfolding experiments: at pH 7, the unfolding temperature T_m is 38.4 °C and 46.2 °C for MBD and NTD-MBD-ID, respectively, and the unfolding enthalpy ΔH_m is 38 kcal/mol and 46 kcal/mol for MBD and NTD-MBD-ID, respectively [34]. All other domains also have a minor stabilizing effect on MBD [35]. These unfolding experiments were undertaken by following the intrinsic fluorescence emission of the single tryptophan located at the MBD (W104), and consequently they accurately reflect the thermal stability of MBD. This stabilization effect is an intriguing outcome, considering that NTD and ID do not adopt well-defined structures by themselves. There are two main explanations for this phenomenon: (1) NTD and ID, although disordered or with high susceptibility to populate unfolded structures, are able to interact specifically with MBD or unspecifically (e.g., fuzzy interactions) (e.g., through long range polar/electrostatic interactions due to their large content in polar residues), thus increasing the intrinsic stability of MBD against unfolding; and (2) NTD and ID, being disordered and populating a multitude of different unfolded structures, may exert a steric hindrance effect on MBD, promote its compaction, and restrict its early expansion close to the unfolding temperature of the isolated MBD.

It might be speculated that the observed apparent stabilization effect exerted by NTD and ID could be some kind of artifact when determining the apparent stability of MBD due to: (1) using the intrinsic fluorescence of W104 as a local probe just reporting an unfolding process restricted to the vicinity of that residue; or (2) using temperature as a physico-chemical stress for altering the noncovalent interatomic interactions responsible for maintaining the folded MBD structure and, thus, triggering the protein unfolding process. In fact, a slightly destabilizing effect for NTD and ID on MBD was reported before [36], but minor differences in the protein constructs and/or experimental conditions could explain that disagreement. To rule out those possibilities, we have employed differential scanning calorimetry in order to use a thermal unfolding technique providing a global signal with contribution from the entire protein molecule, and we have also monitored the unfolding process by performing chemical denaturations using denaturant concentration as the physico-chemical stress triggering the protein unfolding process. Two MeCP2 constructions have been employed: the isolated MBD, and the MBD together with its two flanking domains, NTD-MBD-ID. The overall conclusion is that both completely disordered domains, NTD and ID, unequivocally stabilize the MBD against thermal and chemical denaturation. Therefore, disorder in proteins may be considered a pervasive feature that plays an important role in many the allosteric control of protein conformation, protein interactions, and protein regulation (modifications, trafficking, degradation, in/activation.). This may be even more important in multidomain IDPs with a complex conformational and multifunctional landscape [13,14].

2. Materials and Methods

2.1. Protein Expression and Purification

Protein variants (MBD and NTD-MBD-ID) were expressed and purified following identical procedures. Plasmids (pET30b) containing both constructions were transformed into BL21 (DE3) Star *E. coli* strain. Starting cultures were grown in 150 mL of LB/kanamycin (50 μg/mL) at 37 °C overnight. Then, 4 L of LB/kanamycin (25 μg/mL) were inoculated (1:100 dilution) and incubated under the same conditions until reaching an optical density (λ = 600 nm) of 0.6. Protein expression was induced by adding isopropyl 1-thio-β-D-galactopyranoside (IPTG) 1 mM at 18 °C overnight.

Cells were sonicated in ice and benzonase (Merck-Millipore, Madrid, Spain) was added (20 U/mL) to remove nucleic acids. Proteins were purified using metal affinity chromatography using a HiTrap TALON column (GE-Healthcare Life Sciences, Barcelona, Spain) with two washing steps: buffer sodium phosphate 50 mM, pH 7, NaCl 300 mM, and buffer sodium phosphate 50 mM, pH 7, NaCl 800 mM. Elution was performed applying an imidazole 10–150 mM elution gradient. Protein purity was evaluated by sodium dodecyl sulfate polyacrylamide gel electrophoresis.

The polyhistidine-tag was removed by processing with GST-tagged PreScission Protease in proteolytic cleavage buffer (Tris-HCl 50 mM, NaCl 150 mM, pH 7.5) at 4 °C for 4 h. Progress of the proteolytic processing was monitored by SDS-PAGE. The protein was further purified with a combination of two affinity chromatographic steps for removing the polyhistidine-tag (HiTrap TALON column) and the GST-tagged PreScission Protease (GST TALON column, from GE-Healthcare Life Sciences, Barcelona, Spain). Purity and homogeneity were evaluated by SDS-PAGE and size-exclusion chromatography. Storage buffer consisted of Tris 50 mM pH 7.0 and pooled samples were kept at −80 °C. The identity of all proteins was checked by mass spectrometry (4800plus MALDI-TOF/MS, from Applied Biosystems-Thermo Fisher scientific, Waltham, MA, USA). Potential DNA contamination was always estimated by UV absorption 260/280 ratio. Because a single tryptophan is located in MBD, an extinction coefficient of 11,460 M^{-1} cm^{-1} at 280 nm was employed for the two variants.

2.2. Double-Stranded DNA

HPLC-purified methylated and unmethylated 45-bp single-stranded DNA (ssDNA) oligomers, corresponding to the promoter IV of the mouse brain-derived neurotrophic factor (BDNF) gene [36,37], were purchased from Integrated DNA Technologies (Leuven, Belgium). Two complementary pairs of ssDNA were used for DNA binding assays:

forward unmethylated: 5′-GCCATGCCCTGGAACGGAACTCTCCTAATAAAAG-ATGT ATCATTT-3′;
reverse unmethylated: 5′-AAATGATACATCTTTTATTAGGAGAGTTCCGTTCC-AGGGCA TGGC-3′;
forward mCpG: 5′-GCCATGCCCTGGAA(5-Me)CGGAACTCTCCTAATAAA-AGATGTA TCATTT-3′;
reverse mCpG: 5′-AAATGATACATCTTTTATTAGGAGAGTTC(5-Me)CGTT-CCAGGGCA TGGC-3′.

The ssDNA oligonucleotides were dissolved at a concentration of 0.5 mM and the concentration was assessed by using the extinction coefficient provided by the manufacturer. Complementary ssDNA oligomers were mixed at an equimolar ratio and annealed to obtain 45-bp double-stranded DNA (dsDNA) using a Stratagene Mx3005P qPCR real-time thermal cycler (Agilent Technologies, Santa Clara, CA, USA). The thermal annealing profile consisted of: (1) equilibration at 25 °C for 30 s; (2) heating ramp up to 99 °C; (3) equilibration at 99 °C for 1 min; and (4) 3 h cooling process down to 25 °C at a rate of 1 °C/3 min.

2.3. Thermal Denaturation Assessment by Differential Scanning Calorimetry

Thermal stability of MeCP2 MBD and NTD-MBD-ID was assessed by temperature unfolding monitored by high-precision differential scanning calorimetry (DSC). The partial molar heat capacity of the protein in solution was measured as a function of temperature in an Auto-PEAQ-DSC (MicroCal, Malvern-Panalytical, Malvern, UK). Experiments were performed with a protein solution at a concentration of 20–40 µM in Tris 50 mM pH 7.0, and scanning from 15 to 95 °C at a scan rate of 60 °C/h.

The first approach in DSC experimental analysis consisted of applying a model-free data analysis for discriminating between different possibilities: two-state unfolding, non-two-state unfolding, and oligomer unfolding of the protein. From the thermogram (excess molar heat capacity of the protein as a function of the temperature, ΔC_P (T)), the calorimetric unfolding enthalpy, ΔH_{cal}, the unfolding temperature, T_m, and the maximal unfolding heat capacity, $C_{P,max}$, were estimated, from which it was possible to calculate the van't Hoff enthalpy, ΔH_{vH}:

$$\Delta H_{vH} = \frac{4RT_m^2 C_{P,max}}{\Delta H_{cal}} \tag{1}$$

From the ratio $\Delta H_{\text{vH}}/\Delta H_{\text{cal}}$ different possibilities may arise: (1) if $\Delta H_{\text{vH}}/\Delta H_{\text{cal}} = 1$, the thermogram would reflect a single transition and the protein unfolds according to a two-state model (i.e., the protein contains a single energetic domain); (2) if $\Delta H_{\text{vH}}/\Delta H_{\text{cal}} < 1$, the thermogram would reflect, at least, two (partially) overlapping transitions and the protein unfolds according to a non-two-state model (i.e., the protein contains, at least, two domains which unfold in an independent manner); and (3) if $\Delta H_{\text{vH}}/\Delta H_{\text{cal}} > 1$, the thermogram would reflect an unfolding transition coupled to subunit dissociation (i.e., the protein is oligomeric and unfolds into monomers).

Once the appropriate unfolding model can be selected according to the previous van't Hoff test, the thermogram was analyzed by non-linear regression fitting analysis considering a set of n independent transitions, each characterized by a transition temperature $T_{\text{m,i}}$, an unfolding enthalpy ΔH_i ($T_{\text{m,i}}$), and an unfolding heat capacity $\Delta C_{\text{P,i}}$:

$$\langle \Delta C_{\text{P}}(T) \rangle = \sum_{i=1}^{n} \frac{K_i(T)}{(1+K_i(T))^2} \frac{\Delta H_i(T)^2}{RT^2}$$
$$K_i(T) = \exp\left(-\Delta G_i(T)/RT\right) \tag{2}$$
$$\Delta G_i(T) = \Delta H_i(T_{\text{m,i}})\left(1 - \frac{T}{T_{\text{m,i}}}\right) + \Delta C_{\text{P,i}}\left(T - T_{\text{m,i}} - T\ln\frac{T}{T_{\text{m,i}}}\right)$$

where K_i, ΔH_i, and ΔG_i are the equilibrium constant, the unfolding enthalpy, and the stabilization Gibbs energy for the protein conformational transition i, respectively, R is the ideal gas constant, and T is the absolute temperature.

2.4. Chemical Denaturation Assessment by Fluorescence Spectroscopy

Chemical stability of MeCP2 MBD and NTD-MBD-ID was assessed by fluorescence spectroscopy, using ultra-pure urea (urea crystalline pharma grade, PanReac, Barcelona, Spain) as chaotropic denaturant. The intrinsic fluorescence emission of the single tryptophan in MeCP2 (W104) was monitored inn a thermostated Cary Eclipse fluorescence spectrophotometer (Varian-Agilent, Santa Clara, CA, USA) using a protein concentration of 5 µM, in a 1 cm path-length quartz cuvette (Hellma Analytics, Müllheim, Germany). Experiments were performed in Tris 50 mM pH 7.0, at 20 °C, with temperature controlled by a Peltier unit. Fluorescence emission spectra were recorded from 300 to 500 nm, using an excitation wavelength of 290 nm and a bandwidth of 5 nm, at different urea concentrations ($[D]$ = 0–7 M, with increments of 0.25 M). Samples were equilibrated overnight at room temperature before measurements. Guanidinium hydrochloride was avoided because it is a charged denaturant and would interfere with electrostatic interactions between protein charges and it is known to be a stabilizing molecule at low concentration [38,39]. Considering only the native and the unfolded states of the protein (two-state model) was sufficient for reproducing the denaturation curves, and inclusion of additional intermediate partially unfolded states was detrimental for the goodness of the fitting. Three unfolding-reporting signals were employed for quantitative analysis. The fluorescence emission intensity I at a single wavelength (in the case of MeCP2, 340 nm, i.e., I_{340}), or the intensity ratio at two wavelengths (in the case of MeCP2, I_{330}/I_{350}) as a function of denaturant concentration, which, for an unfolding process involving independent transitions (corresponding to independently unfolding regions or domains), must be analyzed according to this set of equations:

$$I([D]) = \frac{1}{1+K([D])} I_{\text{N}}([D]) + \frac{K([D])}{1+K([D])} I_{\text{U}}([D])$$
$$K([D]) = \exp\left(-\Delta G([D])/RT\right) \tag{3}$$
$$\Delta G([D]) = \Delta G_{\text{w}}([D]) - m[D]$$
$$I_i([D]) = A_i + B_i[D]$$

where K, ΔG, ΔG_{w}, and m are the equilibrium constant, the stabilization Gibbs energy in the presence and the absence of denaturant, and the susceptibility of ΔG to the denaturant concentration, for the conformational transition. The linear extrapolation model [40,41] has been assumed when accounting for the effect of denaturant concentration $[D]$ on the stabilization energy, and the dependence of the intrinsic fluorescence intensity of each protein conformational state with the denaturant concentration was considered to be linear. The average emission energy of the spectrum *<E>* as a function of denaturant concentration, which must be analyzed in a similar way (Equation (3)), was also considered:

$$\langle E \rangle = \frac{\sum_i I_i / \lambda_i}{\sum_i I_i} \tag{4}$$

where I_i is the fluorescence emission intensity in the spectrum at a certain wavelength λ_i. Thus, *<E>* is the spectral average value of the inverse of the wavelength.

The advantage of using the fluorescence intensity is that is more quantitatively rigorous. However, the fluorescence intensity focuses on a certain region of the spectrum, while the average energy adds the possibility of taking into account changes along the entire spectrum for quantifying global spectral changes. Other unfolding observables, such as the wavelength for maximal intensity may lack proportionality with the advance of the unfolding process, may not change significantly upon unfolding, or may make the analysis more complex [42,43].

When needed, methylated and unmethylated dsDNA were added at 6 µM in order to test the ability of the protein to interact with dsDNA and evaluate the extent of the stabilization induced by DNA binding under the same conditions.

2.5. Dynamic Light Scattering (DLS)

Dynamic light scattering measurements were performed in a DynaPro Plate Reader III (Wyatt Technology, Santa Barbara, CA, USA) using a 384-multiwell plate (Aurora Microplates, Whitefish, MT, USA). Urea stock solution (8 M) was serially diluted to obtain concentrations ranging from 0 M to 7 M (steps of 1 M), before the addition of protein to a final concentration of 20 µM. All solutions were filtered using 0.2-µm membranes and protein stocks were centrifuged in microtubes for 2 min at maximum speed to prevent urea precipitation or protein aggregates from interfering with the DLS measurements. For each measurement, 10 acquisitions of 3 s were taken, and the apparent hydrodynamic radius was estimated from the experimental diffusion coefficient, obtained by the cumulant fit of the translational autocorrelation function, assuming a Rayleigh sphere model. Experiments were performed in Tris 50 mM pH 7.0, at 20 °C.

3. Results

3.1. The MBD Is Stabilized against Thermal Unfolding by Its Disordered Flanking Domains

The DSC thermograms showed a single apparent unfolding transition for MBD and NTD-MBD-ID (Figure 1). Experiments performed at 20 and 40 µM protein concentrations provided similar results (similar apparent T_m and unfolding enthalpy values), and therefore the native protein remains monomeric and the unfolding does not trigger oligomerization or aggregation.

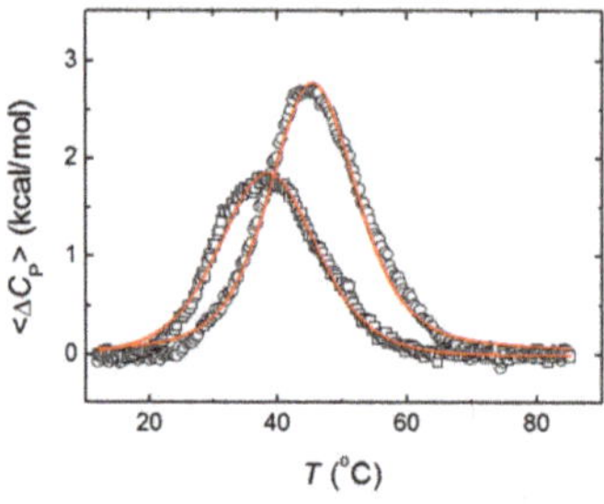

Figure 1. Differential scanning calorimetry thermograms (excess molar heat capacity of the protein as a function of temperature) for MBD (squares) and NTD-MBD-ID (circles). Experiments were done with a concentration of 40 µM, in Tris 50 mM, pH 7. Non-linear fittings are shown (red), according to a single transition model (for MBD) and a two independent transitions model (for NTD-MBD-ID).

The model-free van't Hoff analysis indicated that MBD unfolds as a single transition, whereas NTD-MBD-ID unfolds through two independent transitions, because the enthalpies ratio is close to 1 for MBD, but rather lower than 1 for NTD-MBD-ID (Table 1). The thermal unfolding monitored by fluorescence (using the single tryptophan W104 as an intrinsic probe) was previously analyzed according to a single transition (two-state model) and provided unfolding parameters in reasonable agreement with those reported now: transition temperature of 38.4 and 46.2 °C for MBD and NTD-MBD-ID, respectively, and unfolding enthalpy of 29 and 37 kcal/mol for MBD and NTD-MBD-ID, respectively. Thus, it seems that the presence of the two disordered flanking domains increased the stability of MBD and the unfolding process of the longer construct occurred with unfolding intermediates.

Table 1. Parameters associated with the van 't Hoff test (or calorimetric two-state test).

Protein Construction	T_m (°C)	ΔH_{cal} (kcal/mol)	C_{Pmax} (kcal/K·mol)	ΔH_{vH} (kcal/mol)	Ratio
MBD	37.8	36	1.85	39	1.08
NTD-MBD-ID	44.8	52	2.73	42	0.81

According to these results, the MBD unfolding was analyzed with a model considering a single transition and that for NTD-MBD-ID was analyzed with a model considering two independent transitions (Figure 1 and Table 2).

Table 2. Thermal unfolding parameters for MBD and NTD-MBD-ID at pH 7.

Protein Construction	T_{m1} (°C)	ΔH_1 (kcal/mol)	T_{m2} (°C)	ΔH_2 (kcal/mol)	Sqrt(RSS) (kcal/mol)
MBD	37.4 ± 0.2	38 ± 1	–	–	1.0603
NTD-MBD-ID	45.0 ± 0.2	47 ± 1	–	–	2.0606
	41.7 ± 0.2	11 ± 1	45.0 ± 0.2	46 ± 1	1.5540

Sqrt(RSS): square root of the residual sum of squares.

Although the unfolding of NTD-MBD-ID could be reasonably fitted with a single transition model, the residuals sum of squares (RSS) was smaller for the model with two transitions. More rigorously, the parametric F-test indicated that the model considering two transitions is statistically more appropriate ($F = 50.40 > F_{3213}$ ($\alpha = 0.05$) = 2.65). To observe other potential differences between MBD and NTD-MBD-ID, and to compare stabilities under different physico-chemical stresses, fluorescence chemical denaturations using urea as chaotropic denaturant were performed.

3.2. The MBD Is Stabilized against Chemical Unfolding by Its Disordered Flanking Domains

Because there is a single tryptophan residue (W104) in the whole MeCP2 sequence, and it is located in the MBD, measuring the intrinsic fluorescence of this residue is a good indicator of the folding state of MBD in both constructions. In the absence of denaturant, both protein constructs showed non-symmetrical, bell-shaped spectra, with a maximum around 340 nm indicating that W104 was not exposed to the solvent (Figure 2). As the concentration of denaturant increased, the intensity of emission was dramatically reduced (quenching) and red-shifted, with a maximum near 350 nm, as the fraction of unfolded protein increased, indicating that W104 was exposed to the solvent.

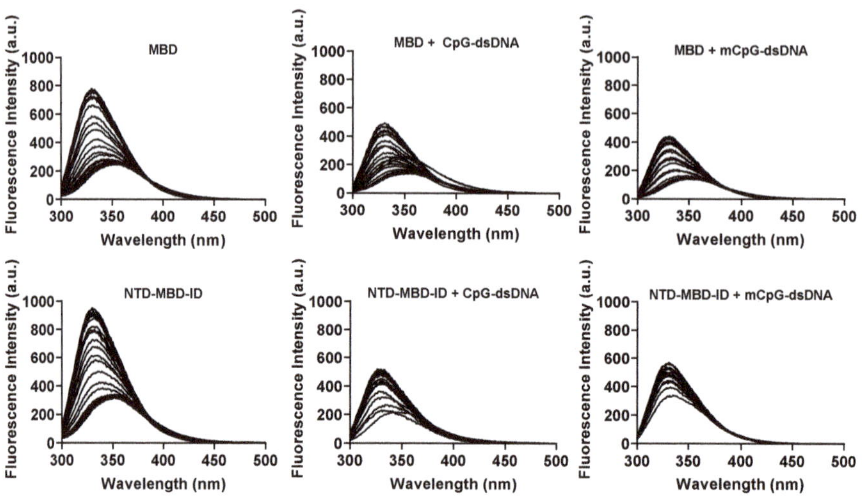

Figure 2. Raw fluorescence spectra for MBD (top) and NTD-MBD-ID (bottom) at different urea concentrations ([D] = 0–7 M) in the absence of dsDNA (**left**), in the presence of unmethylated CpG-dsDNA (**middle**), and methylated mCpG-dsDNA (**right**).

Protein stabilization upon dsDNA was also determined by fluorescence chemical denaturation assays. When dsDNA was present, the emitted fluorescence intensity was considerably diminished due to light absorption by dsDNA at these wavelengths. Raw spectra were processed in order to calculate the spectral average energy at each experimental condition, and the unfolding traces were constructed (Figure 3).

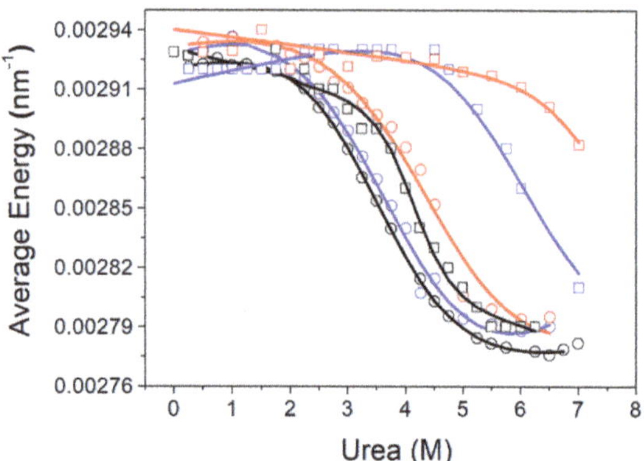

Figure 3. Chemical unfolding curves for MBD (circles) and NTD-MBD-ID (squares). The spectral average energy of the protein was represented as a function of the urea concentration: free protein (black), protein bound to unmethylated CpG-dsDNA (blue), and methylated mCpG-dsDNA (red).

The disordered flanking domains exerted a subtle stabilizing effect on MBD, as seen in the chemical unfolding curves monitoring the spectral average energy (Figure 3 and Table 3). The binding of dsDNA always produced stabilization of MBD against chemical denaturation (Figure 3 and Table 3).

Table 3. Thermal unfolding parameters for MBD and NTD-MBD-ID at pH 7, estimated by by analyzing the denaturant dependence of the spectral average energy.

Protein Construction	DNA	ΔG_w (kcal/mol)	m (kcal/mol·M)	$[D]_{1/2}$ (M)
MBD	–	2.5	0.77	3.3
	CpG-dsDNA	2.7	0.68	3.9
	mCpG-dsDNA	3.2	0.67	4.8
NTD-MBD-ID	–	2.8	0.70	4.0
	CpG-dsDNA	4.2	0.72	5.9
	mCpG-dsDNA	7.6	0.97	7.8

$[D]_{1/2}$ is the half-unfolding denaturant concentration. The stabilization Gibbs energy in the absence of denaturant is related to $[D]_{1/2}$: $\Delta G_w = m \, [D]_{1/2}$.

The stabilizing effect was dramatically increased when the dsDNA was present. It must be born in mind that the presence of ID not only increases 400-fold the dsDNA binding affinity (from micromolar affinity to nanomolar affinity), but also provides an additional binding dsDNA site with micromolar affinity (the extent of the stabilization effect depends on the stoichiometry of the interaction and the binding affinity, among other factors) [34]. The interaction with mCpG-dsDNA was strikingly stabilizing for the protein. Indeed, even at high urea concentrations (i.e., $[D] > 6$ M), a large fraction of the protein (> 50%) seemed to remain folded. Unfolding traces could be satisfactorily analyzed employing an unfolding model with a single transition, and the unfolding parameters were estimated by non-linear least-squares regression analysis (Table 3).

For comparison, the spectral series shown in Figure 2 were also analyzed focusing on the fluorescence intensity at a single wavelength (340 nm), as shown in Figure 4 and Table 4, and the intensity ratio at two wavelengths (to reduce potential uncertainty and variability due to the protein concentration), as shown in Figure 5 and Table 5.

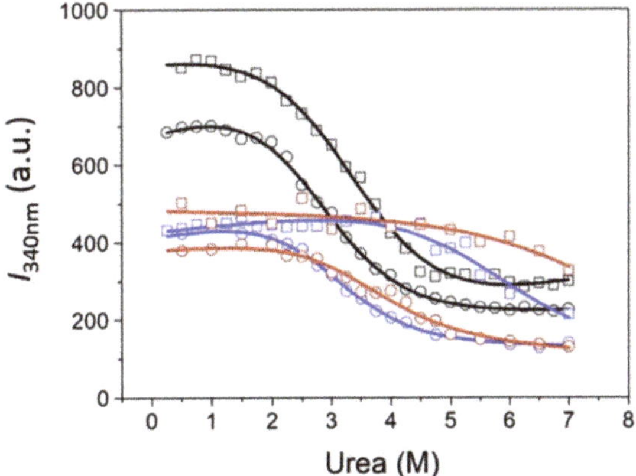

Figure 4. Chemical unfolding curves for MBD (circles) and NTD-MBD-ID (squares). The fluorescence intensity of the protein at 340nm was represented as a function of the urea concentration: free protein (black), protein bound to unmethylated CpG-dsDNA (blue), and methylated mCpG-dsDNA (red).

Table 4. Thermal unfolding parameters for MBD and NTD-MBD-ID at pH 7, estimated by analyzing the denaturant dependence of the fluorescence intensity at 340 nm.

Protein Construction	DNA	ΔG_w (kcal/mol)	m (kcal/mol·M)	$[D]_{1/2}$ (M)
MBD	–	2.3	0.84	2.7
	CpG-dsDNA	2.7	0.92	2.9
	mCpG-dsDNA	2.9	0.83	3.5
NTD-MBD-ID	–	2.7	0.78	3.4
	CpG-dsDNA	3.6	0.64	5.6
	mCpG-dsDNA	3.8	0.49	7.7

$[D]_{1/2}$ is the half-unfolding denaturant concentration. The stabilization Gibbs energy in the absence of denaturant is related to $[D]_{1/2}$: $\Delta G_w = m\,[D]_{1/2}$.

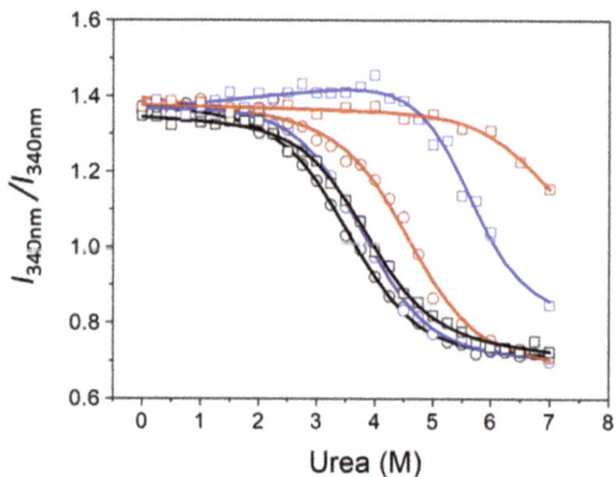

Figure 5. Chemical unfolding curves for MBD (circles) and NTD-MBD-ID (squares). The ratio of fluorescence intensities at 330 and 350 nm was represented as a function of the urea concentration: free protein (black), protein bound to unmethylated CpG-dsDNA (blue), and methylated mCpG-dsDNA (red).

Table 5. Thermal unfolding parameters for MBD and NTD-MBD-ID at pH 7, estimated by analyzing the denaturant dependence of the fluorescence intensity ratio I_{330}/I_{350}.

Protein Construction	DNA	ΔG_w (kcal/mol)	m (kcal/mol·M)	$[D]_{1/2}$ (M)
MBD	–	3.6	1.02	3.5
	CpG-dsDNA	3.9	1.04	3.7
	mCpG-dsDNA	4.8	1.06	4.6
NTD-MBD-ID	–	4.1	1.09	3.8
	CpG-dsDNA	6.9	1.24	5.6
	mCpG-dsDNA	7.7	1.10	7.0

$[D]_{1/2}$ is the half-unfolding denaturant concentration. The stabilization Gibbs energy in the absence of denaturant is related to $[D]_{1/2}$: $\Delta G_w = m\,[D]_{1/2}$.

3.3. MBD Molecular Size Is Highly Susceptible to the Presence of the Flanking Domains

The apparent hydrodynamic radius of the two constructs, MBD and NTD-MBD-ID, was measured under different conditions: different urea concentrations and in the absence/presence of dsDNA (Figure 6 and Table 6). The size histograms showed average and standard deviation values that were modulated by urea concentration (Supplementary

Figure S2). In addition, we could observe a similar effect from unmethylated and methylated dsDNA.

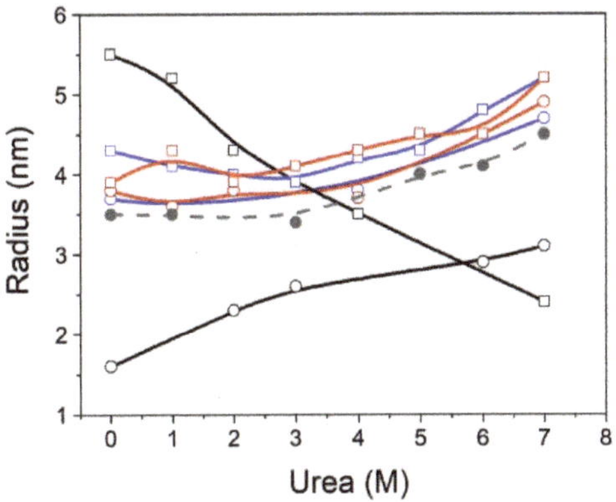

Figure 6. Apparent hydrodynamic radius of MBD and NTD-MBD-ID measured by DLS of MBD (empty circles) and NTD-MBD-ID (squares) in the absence of DNA (black), in the presence of unmethylated CpG-dsDNA (blue), and the presence of methylated mCpG-dsDNA (red). Control measurements of dsDNA are also shown (gray circles).

Table 6. Hydrodynamic radius of the two protein constructs in complex with dsDNA estimated by dynamic light scattering.

Protein Construction	DNA	Radius (nm)
	–	1.6
MBD	CpG-dsDNA	3.7
	mCpG-dsDNA	3.8
	–	5.5
NTD-MBD-ID	CpG-dsDNA	4.3
	mCpG-dsDNA	3.9
–	CpG-dsDNA	3.5

In the absence of urea, MBD showed a hydrodynamic radius close to that predicted from its molecular weight. As the concentration of urea increased, the apparent hydrodynamic radius of MBD increased, as expected for an unfolding process, but the apparent hydrodynamic radius of NTD-MBD-ID decreased. This unexpected result may be related to the large proportion of disorder, the large proportion of charged residues, and the increase in dielectric constant of the solvent when urea concentration increases, as discussed later. The interpretation of the low susceptibility and small increase of the apparent hydrodynamic radius of the two constructs to the concentration of urea in the presence of dsDNA is difficult, because it must be a combination of several effects: initial stabilization of protein by dsDNA interaction is weakened because of the reduction in the dielectric constant of the solvent (weakened polar/electrostatic interactions between protein and dsDNA) and the preferential interaction of urea with the protein residues displacing water molecules, leading to dsDNA dissociation and protein destabilization and unfolding.

4. Discussion

The impact of disordered regions on the stability and functional features of well-folded regions in proteins remains as an important and elusive matter intimately connected with protein function regulation and allosteric control. There are some cases where this issue gets especially important. For example, the stabilization induced by intrinsically disordered regions in the HIV-1 Rev protein has been reported [21]. In another example, stability changes have been described for nucleoplasmin depending on the length of the disordered tail in each subunit of the pentameric protein: a fifty-residue C-terminal deletion mutant showed lower thermal stability, whereas an eighty C-terminal deletion mutant showed higher thermal stability than the full-length protein [22]. In the case of MeCP2, it has been previously outlined that the two flanking domains of MBD in MeCP2 (NTD and ID) increase MBD stability and dsDNA binding affinity. This is not a simple issue, because those two domains are fully disordered and MBD is also 40% disordered, approximately. The comparison of the circular dichroism spectra for both constructions reflected a lower level of structural order for NTD-MBD-ID (Supplementary Figure S1). The stabilization effect was observed through fluorescence thermal denaturations, and two reasonable objections could be pointed out: (1) the lack of a signal reporting global effects in the protein, because fluorescence intensity only reports the local effect in the surroundings of the single tryptophan in MBD in the thermal denaturations, and (2) the thermal stress driving the unfolding process that could result in an artifactual observation, whereas other unfolding processes (e.g., chemical or pressure denaturation) might provide a somewhat different outcome. In order to rule out these possibilities, we studied the unfolding of MBD and NTD-MBD-ID by differential scanning calorimetry and by chemical denaturation.

Both constructions, MBD and NTD-MBD-ID, showed a single apparent transition in their DSC thermogram. Applying the van 't Hoff (two-state) test, it could be observed that, although MBD seemed to unfold through a single transition, the unfolding of NTD-MBD-ID involved at least two intermediate unfolding states that can be significantly populated at moderate temperatures. Thus, the conformational landscape of NTD-MBD-ID is more complex than expected, and its unfolding cooperativity is lower than that of MBD. The two unfolding transitions observed in NTD-MBD-ID might correspond to two independent transitions within MBD (the presence of the flanking domains decouples two regions and lowers the unfolding cooperativity of MBD) or might reflect a transition in NTD or ID corresponding to a region that adopts a folded conformation when accompanying the MBD and undergoes unfolding.

The stabilization effect exerted by NTD and ID was evident from the comparison of the apparent transition temperatures for both constructs determined by DSC, namely 37.4 °C for MBD and 45 °C for NTD-MBD-ID, which are fairly similar to those determined by fluorescence thermal denaturations (38.4 °C for MBD and 46.2 °C for NTD-MBD-ID). However, a striking phenomenon can be noticed with a further analysis. From the unfolding parameters reported in Table 2, the molar fractions of the relevant conformational states for MBD and NTD-MBD-ID at any temperature can be outlined (Figure 7). It can be observed that at 20 °C, 96% of MBD remains in its native conformation, and the temperature for the unfolded state to be 50% populated was 37.4 °C (the transition temperature). From that, a stabilization Gibbs energy at 20 °C of 1.9 kcal/mol could be estimated, in fair agreement with the 1.4 kcal/mol estimated by thermal fluorescence denaturations. However, although the completely unfolded state of NTD-MBD-ID reaches a 50% population at 50.7 °C, the native state is populated only 75% at 20 °C with a corresponding stabilization Gibbs energy of just 0.6 kcal/mol, because of the coexistence of partially unfolded states. Therefore, it is obvious that NTD-MBD-ID requires higher temperatures for complete unfolding compared to MBD, but at low temperature the relevant stability (the stabilization energy gap connecting the native state and the first partially unfolded state) [44] is lower than that of MBD (0.6 kcal/mol for NTD-MBD-ID and 1.9 kcal/mol for MBD), and the integrity of the NTD-MBD-ID native state is compromised to a larger extent at low temperatures. For example, at 37.4 °C (a temperature at which the stabilization Gibbs energy for MBD is zero and both native state and unfolded state are 50% populated), the stabilization Gibbs energy of NTD-MBD-ID is –0.7 kcal/mol and the native state is only 46% populated.

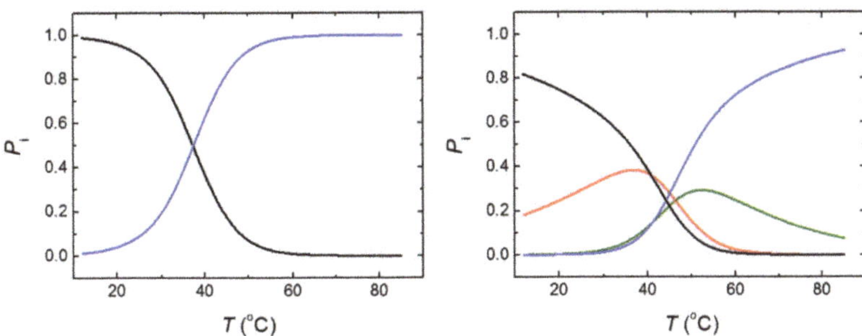

Figure 7. Molar fraction or population (P_i) of the different conformational states: native state (black), first intermediate state (red), second intermediate state (green), and fully unfolded state (blue), for MBD (left) and for NTD-MBD-ID (right). Populations were calculated with the parameters shown in Table 2.

The lack of correlation between the apparent T_m's and the stabilization Gibbs energy at low temperature is an indication of the caveats associated with using only T_m's for assessing structural stability. A rigorous stability assessment must involve the determination of stabilization Gibbs energies at relevant conditions. This problem parallels with the often-observed lack of correlation between the half-unfolding denaturant concentration $[D]_{1/2}$ and the stabilization Gibbs energy ΔG_w, as discussed below.

Temperature and chemical denaturation modulate and alter intramolecular noncovalent interactions (e.g., hydrogen bonds, electrostatic interactions and salt bridges, hydrophobic interactions) by different mechanisms. Temperature alters the thermal motion, lowers the barrier to overcome the stabilizing interatomic energies, and reduces the dielectric constant, with an overall destabilizing effect [45]. Urea increases the dielectric constant (higher than water) [46], diminishing the strength of polar/charge interactions, and interacts specifically with the protein backbone and displaces water molecules by preferential interaction lowering the hydration capability of water, destabilizing the native state, and shifting the conformational equilibrium towards (partially and completely) unfolded states, with an overall destabilizing effect. The different effect of temperature on the dielectric constant (and therefore, on the strength of electrostatic interactions) is relevant, since IDPs such as MeCP2 contain a considerable percentage of polar/charged residues, and specific and unspecific electrostatic interactions may be established intramolecularly. Thus, electrostatic interactions, either specific direct interactions or long-range unspecific interactions, surely play an important role mediating the reciprocal effect between MBD and both flanking domains NTD and ID. An interesting observation arises from the fact that the stabilization effect is considerably diminished at pH 7 when the ionic strength is increased: at NaCl 150 mM, the unfolding temperature T_m is 46.4 °C and 49.8 °C for MBD and NTD-MBD-ID, respectively, and the unfolding enthalpy ΔH_m is 32 kcal/mol and 38 kcal/mol for MBD and NTD-MBD-ID, respectively. IDPs and IDRs are rich in polar/charged residues, and the ionic screening on electrostatic charges will affect both the specific and the unspecific mechanisms of interdependence between folded and unstructured domains.

The chemical denaturation of MBD and NTD-MBD-ID followed by fluorescence seem to occur as a single transition. From the experimental data, a clear stabilization effect of NTD and ID can be observed from the apparent half-unfolding denaturant concentration: $[D]_{1/2}$ is 3.30 M for MBD and 3.95 for NTD-MBD-ID. However, when calculating the stabilization Gibbs energy ΔG_w, the difference is not as clear, and the changes observed in $[D]_{1/2}$ be obscured in ΔG_w because of the m values. A similar phenomenon occurs when calculating stabilization Gibbs energies at low temperature from the T_m, in which the unfolding heat capacity and enthalpy values are critical for a correct extrapolation. Therefore, estimating stabilities just focusing on T_m or $[D]_{1/2}$ is risky and limited; stabilization energies at a reference experimental condition (e.g., 20 °C and absence of denaturant) must be determined when assessing structural stability.

As previously reported, focusing on different observable signals may result in considerable differences in estimated stability parameters from spectroscopic unfolding curves [42]. In the case of MeCP2, it seems that the analysis based on the fluorescence intensity at 340 nm underestimates the stabilization energies for NTD-MBD-ID in the presence of dsDNA, but the corresponding unfolding traces are ill-defined in the post-transition region because of the high stability. Moreover, the analysis based on the fluorescence intensity ratio seems to overestimate stabilization energies for most of the cases. In this respect, it must be born in mind that the intensity ratio might lack proportionality with the advance of the unfolding process [47,48]. Although the absolute values of stabilization energies do not completely agree, the differences in stabilization energies between protein constructs are rather independent of the signal employed; in particular, the presence of NTD and ID increases the stability of MBD by 0.4 kcal/mol. Thus, even considering the present limitations, the reported data confirm the stabilizing effect of the disordered domains, NTD and ID, on the stability of MBD.

The previously reported stabilization Gibbs energies against thermal denaturation [34] are similar to those determined from chemical denaturation, but they are not in full agreement. At 20 °C, MBD showed a stabilization energy of 1.4 kcal/mol determined from fluorescence thermal denaturations, while it showed a stabilization energy of 2.5 kcal/mol determined by chemical denaturation. Moreover, NTD-MBD-ID showed a stabilization energy of 2.5 kcal/mol determined from fluorescence thermal denaturations, while it showed a stabilization energy of 2.8 determined by chemical denaturation. Of course, those quantities do not necessarily must be in agreement, because, although the initial state in both unfolding processes is the same (the native state), the final state may be different: the unfolded state after thermal denaturation may be structurally different from the unfolded state after chemical denaturation [49,50]. The same occurs with the unfolding process itself, which may proceed through different routes and involving different intermediate states depending on whether temperature or denaturant concentration drives the unfolding process.

Regarding the DLS data, MBD showed a hydrodynamic radius similar to that predicted by size-scaling based on molecular weight, but NTD-MBD-ID showed a much larger hydrodynamic radius than the predicted one [51,52]. If MBD were completely folded or unfolded, the hydrodynamic radius would be expected to be around 1.9 nm or 2.9 nm, respectively, and the observed radius is 1.6 nm; therefore, MBD must contain a considerable percentage of folded structure, as expected. However, NTD-MBD-ID showed an abnormally large radius (5.5 nm), much larger than that predicted even if fully unfolded. If NTD-MBD-ID were completely folded or unfolded, the hydrodynamic radius would be expected to be around 2.5 nm or 4.6 nm. Therefore, assuming the agreement is reasonable (considering the experimental uncertainties and important approximations applied when estimating the apparent hydrodynamic radius by DLS, as well as the uncertainties associated with estimating the size from the protein molecular weight), NTD-MBD-ID must contain a very large proportion of disorder. It is interesting to point out that the completely unfolded NTD is expected to have a hydrodynamic radius of around 2.6 nm, and the completely unfolded ID is expected to have a hydrodynamic radius of around 1.8 nm. Figure 8 shows the hydrophobicity profile of NTD-MBD-ID, which shows two hydrophobic regions, at the end of NTD and the second half of MBD. Disorder predictions (by DISOPRED3 and IUPRED2 algorithms [53,54]) reveal that structural order is confined to most of MBD, and NTD and ID may remain mostly unstructured.

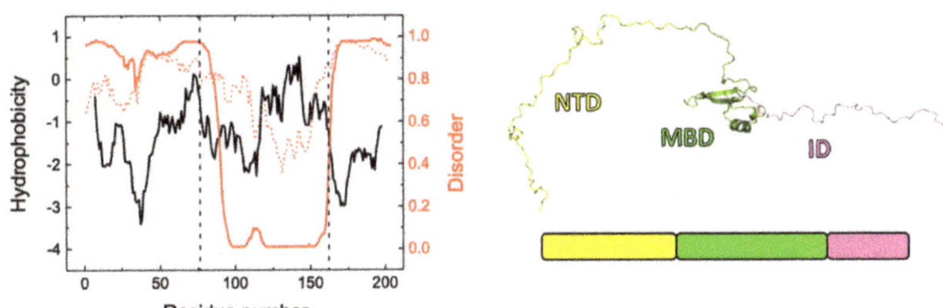

Figure 8. (Left) Hydrophobicity (Kyte & Doolittle scale) plot for NTD-MBD-ID (black) calculated using Expasy ProtScale tool (https://web.expasy.org/protscale/, accessed on 1 July 2021), and disorder prediction plots according to DISOPRED3 (red, continuous line) and IUPRED2 (red, dotted line). The separation between the three domains, NTD, MBD, and ID, are indicated by the vertical dashed lines. (Right) One of the possible conformational states of NTD-MBD-ID, showing the two disordered domains, NTD (yellow) and ID (pink), flanking the folded domain (MBD).

The influence of NTD and ID on MBD may be twofold: (1) unspecific, through steric constraining and long-range electrostatic interactions, and (2) specific, through direct interaction with certain regions in the MBD. Both phenomena would result in spatial restriction of the MBD and additional stabilization, hindering its ability to unfold at moderate temperatures and denaturant concentrations. The interdependence between the different domains in MeCP2 was observed before. Acrylamide collisional quenching, which reports the solvent accessibility of the single tryptophan in MeCP2 (located in MBD), revealed the structural coupling between MBD and all other domains [36]. The structure of the MBD was influenced other domains leading to the shielding of W104 from solvent exposure, with all domains contributing, especially NTD, ID, and TRD.

If the influence of the disordered domains on MBD were mostly due to electrostatic interactions, the ionic strength would strongly modulate the structural stability of MBD. It was reported that increasing the intrinsic ionic strength (NaCl 150 mM) decreases the stabilization effect of the disordered flanking domains (ΔT_m = 7.8 °C with zero ionic strength compared to ΔT_m = 3.4 °C with NaCl 150 mM). However, there is still a substantial stabilization effect even though electrostatic interactions are screened. This suggests the steric/conformational effects are important for the stabilization effect on MBD.

Regarding the experiments with dsDNA, the same phenomena previously observed by fluorescence thermal denaturations were also observed by chemical denaturations: the stabilization effect induced by dsDNA, the larger stabilization effect of dsDNA on NTD-MBD-ID compared to that on MBD, and the larger stabilization exerted by methylated mCpG-dsDNA compared to unmethylated CpG-dsDNA [34]. Still, there is no clue about the underlying mechanism by which such a moderate difference in the in vitro affinity of MeCP2 for dsDNA between the two methylation states (less than five-fold difference in affinity [34]) is further translated to a considerable difference in stabilization upon binding. This hierarchy we have seen in the unfolding experiments is more coherent with the in vivo situation, where MeCP2 acts as a consistent reader of methylation in chromatin.

5. Conclusions

In multidomain proteins, there must be a reciprocal influence of the different constitutive domains, an effect that may be qualitatively and quantitatively different in IDPs because of the larger dynamic sampling of structurally diverse conformational states. Considerable flexibility and a highly dynamic conformational landscape, large susceptibility to environmental conditions, and potential interactions with partners might cause this interdependence to be conditional and dependent on many intervening factors, as well as provide an additional regulation level for protein conformation and function (i.e., allosteric control). MeCP2 is a multidomain IDP in which MBD, the main domain responsible for

DNA recognition, is substantially influenced by its flanking disordered domains. Not only does NTD-MBD-ID differ functionally from MBD, but NTD-MBD-ID also shows differential structural features compared to MBD: (1) conformational landscape with higher complexity, where partially unfolded states may be (functionally) relevant; (2) lower unfolding cooperativity due to the coexistence of partially unfolded intermediate states; and (3) lower relevant stability than MBD but higher overall stability at moderate temperature. In addition, MBD mutations associated with Rett syndrome have different impacts depending on the molecular context, the isolated MBD or the NTD-MBD-ID construct [55]. The application of other experimental techniques (e.g., nuclear magnetic resonance, small-angle X-ray scattering) and computational studies applying molecular dynamic simulations with MBD and NTD-MBD-ID remain very challenging considering the disordered nature of the protein, which might reveal additional effects on MBD from NTD and ID. Moreover, it is very likely that the complexity of the behavior of MeCP2 will be higher for the full-length protein containing the six structural domains.

Supplementary Materials: The following are available online at https://www.mdpi.com/article/10.3390/biom11081216/s1, Figure S1: Circular dichroism, Figure S2: DLS size histograms.

Author Contributions: Conceptualization, M.E., O.A. and A.V.-C.; methodology, O.A. and A.V.-C.; software, O.A. and A.V.-C.; validation, D.O.-A., R.C.-G., S.V., O.C.J.-T., M.E., O.A. and A.V.-C.; formal analysis, D.O.-A., R.C.-G., S.V., O.C.J.-T., M.E., O.A. and A.V.-C.; investigation, D.O.-A., R.C.-G., S.V. and O.C.J.-T.; resources, M.E., O.A. and A.V.-C.; data curation, D.O.-A. and R.C.-G.; writing—original draft preparation, D.O.-A., O.A. and A.V.-C.; writing—review and editing, D.O.-A., R.C.-G., S.V., O.C.J.-T., M.E., O.A. and A.V.-C.; visualization, D.O.-A., R.C.-G., S.V., O.C.J.-T., M.E., O.A. and A.V.-C.; supervision, M.E., O.A. and A.V.-C.; project administration, O.A. and A.V.-C.; funding acquisition, O.A. and A.V.-C. All authors have read and agreed to the published version of the manuscript.

Funding: This research was funded by the Spanish Ministry of Economy and Competitiveness and European ERDF Funds (MCIU/AEI/FEDER, EU) (BFU2016-78232-P to A.V.C.; BES-2017-080739 to D.O.A.); Miguel Servet Program from Instituto de Salud Carlos III (CPII13/00017 to O.A.); Fondo de Investigaciones Sanitarias from Instituto de Salud Carlos III and European Union (ERDF/ESF, "Investing in your future") (PI15/00663 and PI18/00349 to O.A.); Diputación General de Aragón (Protein Targets and Bioactive Compounds Group E45_20R to A.V.C. and Digestive Pathology Group B25_20R to O.A.); and the Centro de Investigación Biomédica en Red en Enfermedades Hepáticas y Digestivas (CIBERehd).

Institutional Review Board Statement: Not applicable.

Informed Consent Statement: Not applicable.

Data Availability Statement: The data presented in this study are available upon reasonable request from the corresponding author.

Acknowledgments: The authors acknowledge the help of J.J. Galano-Frutos and J. Sancho to build the structural file for the NTD-MBD-ID construction.

Conflicts of Interest: The authors declare no conflict of interest. The funders had no role in the design of the study; in the collection, analyses, or interpretation of data; in the writing of the manuscript; or in the decision to publish the results.

References

1. Dyson, H.J. Making sense of intrinsically disordered proteins. *Biophys. J.* **2016**, *110*, 1013–1016. [CrossRef] [PubMed]
2. Uversky, V.N. Intrinsically disordered proteins and their "mysterious" (meta)physics. *Front. Phys.* **2019**, *7*, 10. [CrossRef]
3. Hausrath, A.C.; Kingston, R.L. Conditionally disordered proteins: Bringing the environment back into the fold. *Cell. Mol. Life Sci.* **2017**, *74*, 3149–3162. [CrossRef]
4. Fong, J.H.; Shoemaker, B.A.; Garbuzynskiy, S.O.; Lobanov, M.Y.; Galzitskaya, O.V.; Panchenko, A.R. Intrinsic disorder in protein interactions: Insights from a comprehensive structural analysis. *PLoS Comput. Biol.* **2009**, *5*, e1000316. [CrossRef] [PubMed]
5. Uversky, V.N. Intrinsic disorder-based protein interactions and their modulators. *Curr. Pharm. Des.* **2013**, *19*, 4191–4213. [CrossRef] [PubMed]
6. Prakash, S.; Matouschek, A. Protein unfolding in the cell. *Trends Biochem. Sci.* **2004**, *29*, 593–600. [CrossRef]
7. Fraga, H.; Papaleo, E.; Vega, S.; Velazquez-Campoy, A.; Ventura, S. Zinc induced folding is essential for TIM15 activity as an mtHsp70 chaperone. *Biochim. Biophys. Acta Gen. Subj.* **2013**, *1830*, 2139–2149. [CrossRef] [PubMed]

8.   Darling, A.L.; Uversky, V.N. Intrinsic disorder and posttranslational modifications: The darker side of the biological dark matter. *Front. Genet.* **2018**, *9*, 158. [CrossRef]

9.   Tee, W.-V.; Guarnera, E.; Berezovsky, I.N. Disorder driven allosteric control of protein activity. *Curr. Res. Struct. Biol.* **2020**, *2*, 191–203. [CrossRef] [PubMed]

10.   Berlow, R.B.; Dyson, H.J.; Wright, P.E. Expanding the paradigm: Intrinsically disordered proteins and allosteric regulation. *J. Mol. Biol.* **2018**, *430*, 2309–2320. [CrossRef]

11.   Hilser, V.J.; Thompson, E.B. Intrinsic disorder as a mechanism to optimize allosteric coupling in proteins. *Proc. Natl. Acad. Sci. USA* **2007**, *104*, 8311–8315. [CrossRef]

12.   White, J.T.; Li, J.; Grasso, E.; Wrabl, J.O.; Hilser, V.J. Ensemble allosteric model: Energetic frustration within the intrinsically disordered glucocorticoid receptor. *Philos. Trans. R. Soc. Lond. B Biol. Sci.* **2018**, *373*, 20170175. [CrossRef]

13.   Davey, N.E. The functional importance of structure in unstructured protein regions. *Curr. Opin. Struct. Biol.* **2019**, *56*, 155–163. [CrossRef]

14.   Adams, V.H.; McBryant, S.J.; Wade, P.A.; Woodcock, C.L.; Hansen, J.C. Intrinsic disorder and autonomous domain function in the multifunctional nuclear protein, MeCP2. *J. Biol. Chem.* **2007**, *282*, 15057–15064. [CrossRef] [PubMed]

15.   Schlessinger, A.; Schaefer, C.; Vicedo, E.; Schmidberger, M.; Punta, M.; Rost, B. Protein disorder—A breakthrough invention of evolution? *Curr. Opin. Struct. Biol.* **2011**, *21*, 412–418. [CrossRef] [PubMed]

16.   Schad, E.; Tompa, P.; Hegyi, H. The relationship between proteome size, structural disorder and organism complexity. *Genome Biol.* **2011**, *12*, R120. [CrossRef] [PubMed]

17.   Gao, C.; Ma, C.; Wang, H.; Zhong, H.; Zang, J.; Zhong, R.; He, F.; Yang, D. Intrinsic disorder in protein domains contributes to both organism complexity and clade-specific functions. *Sci. Rep.* **2021**, *11*, 2985. [CrossRef]

18.   Ahrens, J.B.; Nunez-Castilla, J.; Siltberg-Liberles, J. Evolution of intrinsic disorder in eukaryotic proteins. *Cell. Mol. Life Sci.* **2017**, *74*, 3163–3174. [CrossRef]

19.   Franco, G.; Bañuelos, S.; Falces, J.; Muga, A.; Urbaneja, M.A. Thermodynamic characterization of nucleoplasmin unfolding: Interplay between function and stability. *Biochemistry* **2008**, *47*, 7954–7962. [CrossRef]

20.   Taneva, S.G.; Muñoz, I.G.; Franco, G.; Falces, J.; Arregi, I.; Muga, A.; Montoya, G.; Urbaneja, M.A.; Bañuelos, S. Activation of nucleoplasmin, an oligomeric histone chaperone, challenges its stability. *Biochemistry* **2008**, *47*, 13897–13906. [CrossRef]

21.   Faust, O.; Grunhaus, D.; Shimshon, O.; Yavin, E.; Friedler, A. Protein regulation by intrinsically disordered regions: A role for subdomains in the IDR of the HIV-1 Rev protein. *Chem. Bio. Chem.* **2018**, *19*, 1618–1624. [CrossRef]

22.   Hierro, A.; Arizmendi, J.M.; Bañuelos, S.; Prado, A.; Muga, A. Electrostatic interactions at the C-terminal domain of nucleoplasmin modulate its chromatin decondensation activity. *Biochemistry* **2002**, *41*, 6408–6413. [CrossRef]

23.   Ausio, J.; Martinez de Paz, A.; Esteller, M. MeCP2: The long trip from a chromatin protein to neurological disorders. *Trends Mol. Med.* **2014**, *20*, 487–498. [CrossRef]

24.   Guy, J.; Cheval, H.; Selfridge, J.; Bird, A. The role of MeCP2 in the brain. *Annu. Rev. Cell Dev. Biol.* **2011**, *27*, 631–652. [CrossRef]

25.   Tillotson, R.; Bird, A. The molecular basis of MeCP2 function in the brain. *J. Mol. Biol.* **2020**, *432*, 1602–1623. [CrossRef]

26.   Amir, R.E.; Van den Veyver, I.B.; Wan, M.; Tran, C.Q.; Francke, U.; Zoghbi, H.Y. Rett syndrome is caused by mutations in X-linked MECP2, encoding methyl-CpG-binding protein 2. *Nat. Genet.* **1999**, *23*, 185–188. [CrossRef]

27.   Lorenz, J.; Neul, M.D. The relationship of Rett syndrome and MeCP2 disorders to autism. *Dialogues Clin. Neurosci.* **2014**, *14*, 253–262.

28.   Nan, X.; Campoy, F.J.; Bird, A. MeCP2 is a transcriptional repressor with abundant binding sites in genomic chromatin. *Cell* **1997**, *88*, 471–481. [CrossRef]

29.   Hansen, J.C.; Ghosh, R.P.; Woodcock, C.L. Binding of the Rett syndrome protein, MeCP2, to methylated and unmethylated DNA and chromatin. *IUBMB Life* **2010**, *62*, 732–738. [CrossRef] [PubMed]

30.   Nan, X.; Meehan, R.R.; Bird, A. Dissection of the methyl-CpG binding domain from the chromosomal protein MeCP2. *Nucleic Acids Res.* **1993**, *21*, 4886–4892. [CrossRef] [PubMed]

31.   Bellini, E.; Pavesi, G.; Barbiero, I.; Bergo, A.; Chandola, C.; Nawaz, M.S.; Rusconi, L.; Stefanelli, G.; Strollo, M.; Valente, M.M.; et al. MeCP2 post-translational modifications: A mechanism to control its involvement in synaptic plasticity and homeostasis? *Front. Cell. Neurosci.* **2014**, *8*, 236. [CrossRef]

32.   Sheikh, T.I.; de Paz, A.M.; Akhtar, S.; Ausio, J.; Vincent, J.B. MeCP2_E1 N-terminal modifications affect its degradation rate and are disrupted by the Ala2Val Rett mutation. *Hum. Mol. Genet.* **2017**, *26*, 4132–4141. [CrossRef]

33.   Martinez de Paz, A.; Khajavi, L.; Martin, H.; Claveria-Gimeno, R.; Tom Dieck, S.; Cheema, M.S.; Sanchez-Mut, J.V.; Moksa, M.M.; Carles, A.; Brodie, N.I.; et al. MeCP2-E1 isoform is a dynamically expressed, weakly DNA-bound protein with different protein and DNA interactions compared to MeCP2-E2. *Epigenetics Chromatin.* **2019**, *12*, 63. [CrossRef]

34.   Claveria-Gimeno, R.; Lanuza, P.M.; Morales-Chueca, I.; Jorge-Torres, O.C.; Vega, S.; Abian, O.; Esteller, M.; Velazquez-Campoy, A. The intervening domain from MeCP2 enhances the DNA affinity of the methyl binding domain and provides an independent DNA interaction site. *Sci. Rep.* **2017**, *7*, 41635. [CrossRef] [PubMed]

35.   Ortega-Alarcon, D.; Claveria-Gimeno, R.; Vega, S.; Jorge-Torres, O.C.; Esteller, M.; Abian, O.; Velazquez-Campoy, A. Influence of the disordered domain structure of MeCP2 on its structural stability and dsDNA interaction. *Int. J. Biol. Macromol.* **2021**, *175*, 58–66. [CrossRef] [PubMed]

36. Ghosh, R.P.; Nikitina, T.; Horowitz-Scherer, R.A.; Gierasch, L.M.; Uversky, V.N.; Hite, K.; Hansen, J.C.; Woodcock, C.L. Unique physical properties and interactions of the domains of methylated DNA binding protein 2. *Biochemistry* **2010**, *49*, 4395–4410. [CrossRef] [PubMed]
37. Ghosh, R.P.; Horowitz-Scherer, R.A.; Nikitina, T.; Gierasch, L.M.; Woodcock, C.L. Rett syndrome-causing mutations in human MeCP2 result in diverse structural changes that impact folding and DNA interactions. *J. Biol. Chem.* **2008**, *283*, 20523–20534. [CrossRef]
38. Zarrine-Afsar, A.; Mittermaier, A.; Kay, L.E.; Davidson, A.R. Protein stabilization by specific binding of guanidinium to a functional arginine-binding surface on an SH3 domain. *Protein Sci.* **2006**, *15*, 162–170. [CrossRef] [PubMed]
39. Mayr, L.M.; Schmid, F.X. Stabilization of a protein by guanidinium chloride. *Biochemistry* **1993**, *32*, 7994–7998. [CrossRef]
40. Schellman, J.A. Solvent denaturation. *Biopolymers* **1978**, *17*, 1305–1322. [CrossRef]
41. Pace, C.N. Determination and analysis of urea and guanidine hydrochloride denaturation curves. *Methods Enzymol.* **1986**, *131*, 266–280.
42. Eftink, M.R. The use of fluorescence methods to monitor unfolding transitions in proteins. *Biophys. J.* **1994**, *66*, 482–501. [CrossRef]
43. Monsellier, E.; Bedouelle, H. Quantitative measurement of protein stability from unfolding equilibria monitored with the fluorescence maximum wavelength. *Protein Eng. Des. Sel.* **2005**, *18*, 445–456. [CrossRef]
44. Campos, L.A.; Garcia-Mira, M.M.; Godoy-Ruiz, R.; Sanchez-Ruiz, J.M.; Sancho, J. Do proteins always benefit from a stability increase? Relevant and residual stabilisation in a three-state protein by charge optimisation. *J. Mol. Biol.* **2004**, *344*, 223–237. [CrossRef]
45. Wyman, J. Measurements of the dielectric constants of conducting media. *Phys. Rev.* **1930**, *35*, 623. [CrossRef]
46. Wyman, J. Dielectric constants: Ethanol-diethyl ether and urea-water solutions between 0 and 50°. *J. Am. Chem. Soc.* **1933**, *55*, 4116–4121. [CrossRef]
47. Žoldák, G.; Jancura, D.; Sedlák, E. The fluorescence intensities ratio is not a reliable parameter for evaluation of protein unfolding transitions. *Protein Sci.* **2017**, *26*, 1236–1239. [CrossRef]
48. Lopez-Llano, J.; Campos, L.A.; Bueno, M.; Sancho, J. Equilibrium phi-analysis of a molten globule: The 1–149 apoflavodoxin fragment. *J. Mol. Biol.* **2006**, *356*, 354–366. [CrossRef]
49. Lapidus, L.J. Protein unfolding mechanisms and their effects on folding experiments. *F1000Research* **2017**, *6*, 1723. [CrossRef]
50. Narayan, A.; Bhattacharjee, K.; Naganathan, A.N. Thermally versus chemically denatured protein states. *Biochemistry* **2019**, *58*, 2519–2523. [CrossRef]
51. Tyn, M.T.; Gusek, T.W. Prediction of diffusion coefficients of proteins. *Biotechnol. Bioeng.* **1990**, *35*, 327–338. [CrossRef] [PubMed]
52. Armstrong, J.K.; Wenby, R.B.; Meiselman, H.J.; Fisher, T.C. The hydrodynamic radii of macromolecules and their effect on red blood cell aggregation. *Biophys. J.* **2004**, *87*, 4259–4270. [CrossRef] [PubMed]
53. Jones, D.T.; Cozzetto, D. DISOPRED3: Precise disordered region predictions with annotated protein-binding activity. *Bioinformatics* **2015**, *31*, 857–863. [CrossRef]
54. Mészáros, B.; Erdős, G.; Dosztányi, Z. IUPred2A: Context-dependent prediction of protein disorder as a function of redox state and protein binding. *Nucleic Acids Res.* **2018**, *46*, W329–W337. [CrossRef]
55. Ortega-Alarcon, D.; Claveria-Gimeno, R.; Vega, S.; Jorge-Torres, O.C.; Esteller, M.; Abian, O.; Velazquez-Campoy, A. Molecular context-dependent effects induced by Rett syndrome-associated mutations in MeCP2. *Biomolecules* **2020**, *10*, 1533. [CrossRef]

Article

Molecular Dynamics Simulations of Human FOXO3 Reveal Intrinsically Disordered Regions Spread Spatially by Intramolecular Electrostatic Repulsion

Robert O.J. Weinzierl

Department of Life Sciences, Imperial College London, Sir Alexander Fleming Building, Exhibition Road, London SW7 2AZ, UK; r.weinzierl@imperial.ac.uk

Abstract: The human transcription factor FOXO3 (a member of the 'forkhead' family of transcription factors) controls a variety of cellular functions that make it a highly relevant target for intervention in anti-cancer and anti-aging therapies. FOXO3 is a mostly intrinsically disordered protein (IDP). Absence of knowledge of its structural properties outside the DNA-binding domain constitutes a considerable obstacle to a better understanding of structure/function relationships. Here, I present extensive molecular dynamics (MD) simulation data based on implicit solvation models of the entire FOXO3/DNA complex, and accelerated MD simulations under explicit solvent conditions of a central region of particular structural interest (FOXO3$^{120-530}$). A new graphical tool for studying and visualizing the structural diversity of IDPs, the Local Compaction Plot (LCP), is introduced. The simulations confirm the highly disordered nature of FOXO3 and distinguish various degrees of folding propensity. Unexpectedly, two 'linker' regions immediately adjacent to the DNA-binding domain are present in a highly extended conformation. This extended conformation is not due to their amino acid composition, but rather is caused by electrostatic repulsion of the domains connected by the linkers. FOXO3 is thus an IDP present in an unusually extended conformation to facilitate interaction with molecular interaction partners.

Keywords: apoptosis; cancer; longevity; FOXO3; local compaction tool; intrinsically disordered protein; DNA-binding; transcription factor; molecular dynamics simulation; computational biology

Citation: Robert O.J. Weinzierl Molecular Dynamics Simulations of Human FOXO3 Reveal Intrinsically Disordered Regions Spread Spatially by Intramolecular Electrostatic Repulsion. *Biomolecules* **2021**, *11*, 856. https://doi.org/10.3390/biom 11060856

Academic Editors: Thomas R. Caulfield, Simona Maria Monti, Giuseppina De Simone and Emma Langella

Received: 15 April 2021
Accepted: 3 June 2021
Published: 8 June 2021

Publisher's Note: MDPI stays neutral with regard to jurisdictional claims in published maps and institutional affiliations.

1. Introduction

The members of the vertebrate 'forkhead box O' gene family (FOXO1, FOXO3, FOXO4 and FOXO6) act as versatile gene-specific transcription factors that regulate a variety of key cellular processes, such as cellular proliferation, stress response, stem cell maintenance and apoptosis [1]. One of these, FOXO3 (also referred to as FOXO3a, or FKHRL1 in the older literature) is of considerable interest in many therapeutically relevant areas, such as tumor therapy and longevity research [2,3].

Many aspects of FOXO3 activity are controlled by indirect growth factor signaling via phosphatidylinositol 3-kinase (PI3K), Akt and other kinases [4,5]. Post-translational phosphorylation events at specific sites of FOXO3 control the subcellular translocation, association with 14-3-3 proteins [6], DNA-binding properties and transcriptional activity [7]. Additional modifications, such as acetylation and ubiquitination, have also been detected [8]. Like essentially all eukaryotic gene-specific transcription factors, FOXO3 contains a structured DNA-binding domain (DBD) that allows the transcription factor to bind sequence-specifically to target sites in the genome containing a 'forkhead response element' (5′-(G/A)(T/C)(A/C)AA(C/T)A-3′) [9,10]. Many of the genes regulated by FOXO3 carry out anti-proliferative (cell cycle arrest) and pro-apoptotic functions (control of cell death), thus categorizing FOXO3 as a tumor suppressor [5,11,12].

Inspection of the primary sequence of FOXO3 (673 amino acids) reveals a centrally located DBD flanked at both N- and C-termini by sequences predicted to be highly intrinsically disordered (Figure 1A; [13]). The primary amino acid sequence of these intrinsically disordered regions (IDRs) shows a highly uneven distribution of charged and hydrophobic amino acids (Figure 1B). Unlike other intrinsically disordered proteins (IDPs), the primary sequence of FOXO3 is highly conserved among vertebrates (Figure S1).

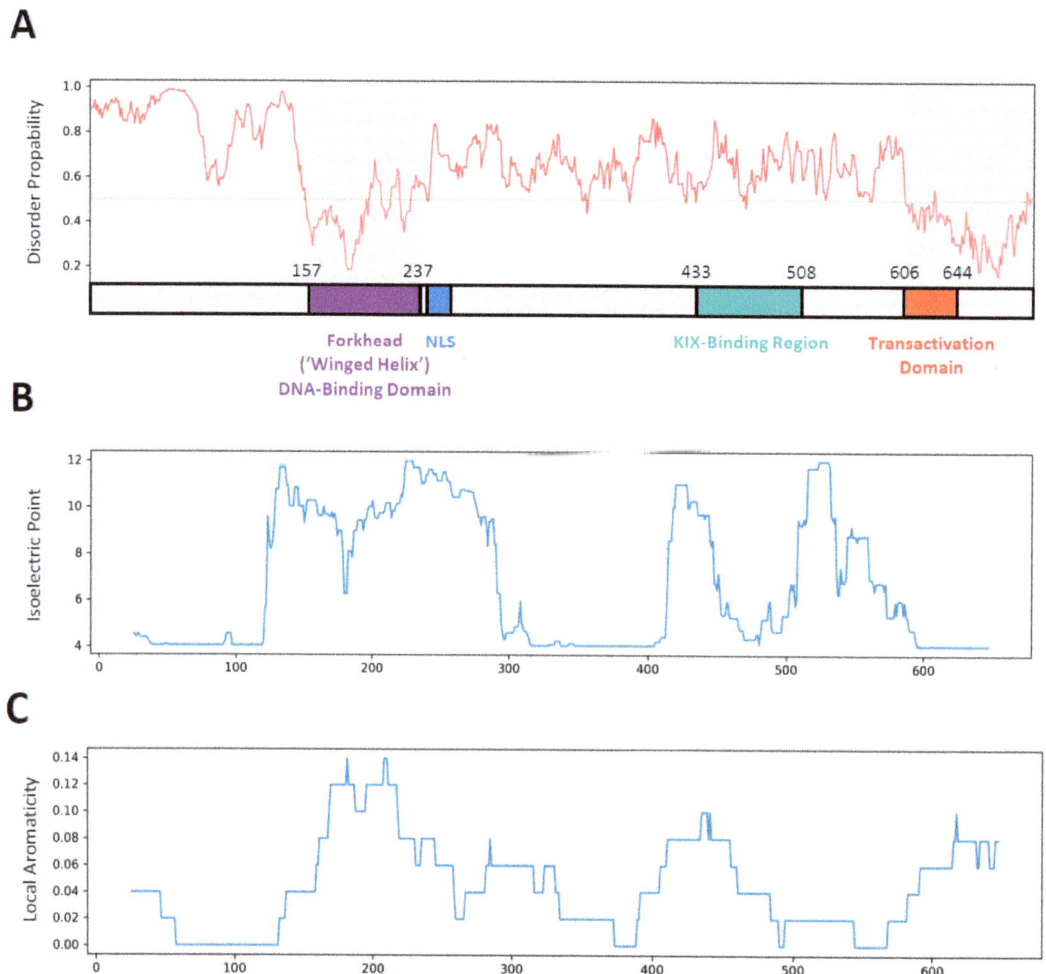

Figure 1. (**A**) Schematic diagram showing the distribution of known functional elements within the primary amino acid sequence of FOXO3 (annotation based on information associated with the NCBI reference sequence NP_963853.1). The start and end positions of several named elements (DNA-Binding Domain (purple), KIX-Binding Region (green) and the transcriptional Transactivation Domain are shown in the diagram). The start and end position of the nuclear localization sequence (NLS, dark blue) are 242 and 259, respectively. The top half of the panel shows the result of an IUPred2A analysis [14] of FOXO3. The red line indicates the probability of residues in that position to participate in a disordered structure. The green line indicates the cut-off point: residues below are considered structured. These values are congruent with the MobiDB-lite scores that predict an overall disorder of ~60% based on multiple criteria and methods (https://mobidb.org/ accessed on 18 May 2021; [15]). (**B**) Local isoelectric point analysis using a sliding window size of 25 residues. The calculated isoelectric value for each window is plotted according to the position of the center of the window. (**C**) Local aromaticity analysis using a sliding window size of 25 residues. The calculated hydrophobicity value (based on the proportion of large aromatic amino acids (Phe, Trp, Tyr) in the window) is plotted according to the position of the center of the window.

Outside the DBD, very little structural information is available. The C-terminal IDR is known to contain the KIX-Binding Region and Transactivation Domain (Figure 1A) that mediate transcriptional activation through their association with the CBP/p300 coactivator complex in a promiscuous manner [16,17]; these regions undergo a transition to α-helices upon binding to the coactivator surface [17].

The important biological roles of FOXO3, combined with the exceptionally high degree of predicted intrinsic disorder, make this transcription factor an interesting target for a more detailed structural characterization by computational simulations to gain further insights into such an unusual protein. Computational molecular dynamics (MD) simulations of intrinsically disordered proteins are challenging because many of the commonly used force-fields and water models have been derived from and optimized for stably folded proteins and are therefore prone to overestimate the structures of IDPs [18–20]. Computational simulation methods are, however, one of the most promising approaches for creating detailed models of IDRs and IDPs [21]; conventional strucural methods are either inapplicable (X-ray crystallography), or provide only population-averaged insights into secondary structure propensity (NMR; for example [22]), or overall molecular shape (SAXS; [21]).

Here, I present a series of atomistic models of full-length FOXO3 bound to DNA based on multiple independent simulations using an implicit MD simulation method, and another series of simulations of the FOXO3-DBD and part of the C-terminal IDR under fully explicit solvation conditions. In a previous study, both simulation conditions gave results that matched experimentally measured parameters well for the disordered N-terminal portion of c-MYC [23], thus showing that such simulation methods are capable of capturing essential structural features of IDPs. The results obtained provide plausible models for a spatially extended structure of FOXO3 when bound to DNA that will provide a better structural understanding to underpin future experimental approaches. A graphical method is introduced that is particularly helpful for visualizing differences in local conformation and folding propensities that will be of general use for analyzing MD trajectories of IDRs and IDPs.

2. Materials and Methods

2.1. Bioinformatic Analysis of Primary Sequence

The primary amino acid sequence of FOXO3 was obtained from the NCBI database (entry NP_963853.1). The sequence was analyzed using the IUPRED2 software [14] running locally. A sliding window analysis (window size 25 amino acids) of isoelectric point and aromaticity employed the Bio.SeqUtils.ProtParam modules from BioPython [24]. Window selection and sliding, data collection and plotting was carried out using custom-written Jupyter notebooks in Python3.

2.2. Creation of the Starting Structure for MD Simulations

The crystal structure of the FOXO3 DNA-binding domain bound to a DNA molecule containing a forkhead response element served as a partial starting structure (PDB# 2UZK). Additional PDB structures of the missing N-terminal (FOXO3$^{1-156}$) and C-terminal (FOXO3$^{254-673}$) IDR sequences were modeled by Markov Chain Monte Carlo simulation [25] and placed in position to link them covalently to the free termini of the DNA-binding domain. The DNA structure was extended using the web application '3DNA 2.0' (http://web.x3dna.org/ accessed on 19 January 2021) with the 'composite' option. This maintains the original conformation of the DNA and contacts with the DNA-binding domain as described in the X-ray structure. The resulting composite structure (Figure S2) was parameterized with the Amber ff14SBonlysc forcefield (in conjunction with OL15 for DNA simulation, igb8 and the mbondi 3 atomic radii set) using the tLEaP program of the AMBER molecular modeling software package [26–28].

2.3. Implicit MD Simulations (gb8_md#1—gb8_md#10)

The simulations were carried out as described previously [23]. Briefly, after minimization, the production simulations were executed on GPUs [29]. Ten independent simulations were run, all starting from the same structure (Figure S2), for a total of 500 nanoseconds each.

2.4. Accelerated Molecular Dynamics of $FOXO^{120-530}$ in TIP4P-D Water

The final structure generated by two of the implicit simulations (gb8_md#1 and gb8_md#2) provided the coordinates of the complete FOXO3 bound to DNA with the flexible linker region (FL#2) C-terminal to the DNA-binding domain in two alternative conformations ('FL#2-A compact' and 'FL#2-A extended'). For preparing the partial structure, the coordinates of residues $FOXO^{1-119}$ were replaced with an acetyl 'cap'. The resulting structures were embedded in a cubic TIP4P-D [18] water box leaving a minimum of 20 Å between the molecule and the edge of the water box. After adjusting the ionic strength to 150 mM NaCl, the final systems containing either 597,606 atoms (Linker Conformation A) or 601,754 atoms (Linker Conformation B) were created. The models were minimized as described previously [23] and subjected to 20 ns of conventional molecular dynamics to obtain stable values for the dihedral and potential energies. These values were used to calculate the parameters for the accelerated MD (aMD) simulations [30], using an α-value of 0.2 as recommended in the Amber manual. Two sets of ten independent production simulations, each lasting 100 ns, were executed on GPUs for each of the two FL#2-A conformations [29].

2.5. Trajectory Visualization and Analysis Methods

The pdb files and MD trajectory data were visualized using Visual Molecular Dynamics (VMD, [31]). For various quantitative analyses (distances, root mean square deviation, secondary structures, etc.), various tools from cpptraj were used [32,33]. The data obtained were analyzed and plotted using custom-written Jupyter Python notebooks.

The pbsa program from the Amber Tools package [27] was employed for the Poisson-Boltzmann PBSA calculations. Standard settings were used except that the ionic strength (istrng) was set to 150 mM, temperature (pbtemp) to 310 K, and the atomic radii read from the prmtop file (radiopt = 0). Volumetric data files for the visualization of electrostatic potential and level set maps were created using the phiout function and graphically represented using VMD [31].

3. Results

3.1. FOXO3 Is Highly Conserved in Evolution, but Represents an Unusually Extensively Disordered IDP

Application of the protein disorder prediction tool IUPred2 [14] to the primary sequence of human FOXO3 suggest the presence of extensively disordered regions located both N- and C-terminally of the structured DNA-binding motif (Figure 1A; [13]). Only the DNA-binding motif and the most C-terminal region, encompassing $FOXO^{600-673}$ are predicted to take up a more folded conformation. The large size of the FOXO3 protein (673 amino acids), in conjunction with the large molecular radius of gyration predicted from its highly disordered nature, precludes atomistic molecular dynamics simulations of the complete molecule under explicitly solvated conditions for technical reasons. I therefore adopted a simulation strategy involving a generalized Born solvation mode (implicit gb8 solvation) MD [25]. The starting model contains extensively elongated N- and C-termini with little or no secondary structure elements to avoid biasing the simulations in any way (Figure S2). The starting model was used to initiate ten independent simulations, each lasting for 500 ns. Analysis of general system variables (root mean square deviation and solvent-accessible surface area) reveals that all simulations take a similar path through conformational space and converge on a series of structures within a relatively narrowly defined phase space (Figure 2A). Visual inspection of the final frames of all ten simulation shows, however, that this does not correspond to a single defined structure but represents a family of related structures with a highly variable composition of secondary structure elements and differently arranged N- and C-terminal sequences (Figure 2B,D; Video S1).

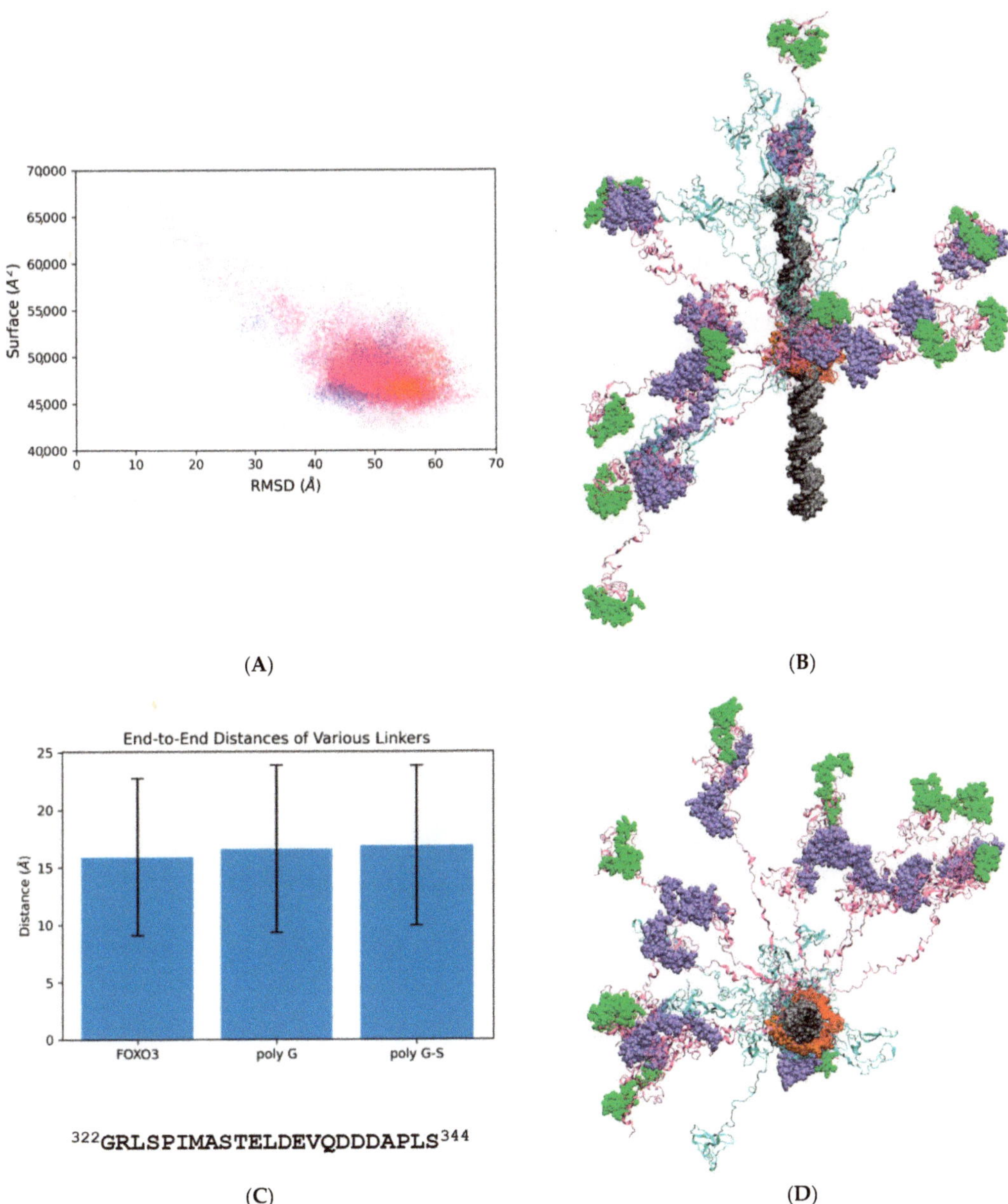

Figure 2. (**A**) Characterization of the molecular dynamics results using global variables (root mean square deviation [RMSD] and solvent accessible surface). The starting structure is located in the top left corner and the dots moving diagonally towards the lower right corner represent intermediate structures formed during the earliest stages of the simulations. All ten simulations finally generate structures that populate a dense space in the lower right corner. Each of the dots (different shades of red, blue, and purple represent different simulations) corresponds to the measurements from a single structure ('snapshot') sampled at 100 picosecond intervals in the trajectories (see Figure S3 for a Gaussian kernel density representation of this result). (**B**) Superimposition of snapshots representing the final structure (at 500 ns) from each of the 10 independent molecular dynamics simulations. The structures were aligned using the stably folded DNA binding motif (FOXO3[157−237]).

The DNA is shown as a silver surface structure. The DNA-binding domain (DBD) is shown as a red cartoon and the nuclear localization sequence in orange van der Waals representation. The intrinsically disordered region (IDR) located N-terminally to the DBD is shown as a cyan cartoon model, and the IDR located C-terminally to the DBD as a pink cartoon model. The KIX-Binding Region is shown as blue van der Waals spheres, and the Transactivation Domain as green spheres. See also Video S1 for a rotating version of the structure for more details and to obtain a three-dimensional understanding of this figure. (**C**) Bar chart of the end-to-end distances of a segment of the flexible linker of FOXO3 (FOXO3$^{322\text{-}344}$, sequence shown below) in comparison to a polypeptide of the same length consisting of glycine residues ('polyG') or alternating glycine-serine residues ('polyG-S') (**D**) Same structure as in (**B**), but viewed down the central axis of the DNA.

Further investigations revealed that in all ten simulations the α-helical elements and β-structure of the 'winged helix' DNA-binding domain remains fully structured as the domain remains bound to its DNA target site in a stable conformation. Both N- and C-terminal IDRs, however, display a variety of variable secondary structure elements (various types of helices, such π-, 3_{10} and α-helices, extended regions (β-sheets) and turns) of variable propensities (Figure 3 and Figure S4). It is of particular interest that in some simulations, a large proportion of extended structures are formed throughout the molecule (for example, gb8_md#1; Figure 3), whereas in other simulations, α- and π-helices predominate as the main secondary structure feature (for example, gb8_md#2 and gb8_md#3; Figure 3). A high proportion of bends and unstructured coils indicate a significantly unfolded state of FOXO3 in both N- and C-terminal IDRs. Once formed, secondary structures frequently only remain folded for a fraction of the simulation time before unfolding (and sometimes refolding) or interconverting to another type of secondary structure. Such fluidity in secondary structure content and stability is expected of an IDP.

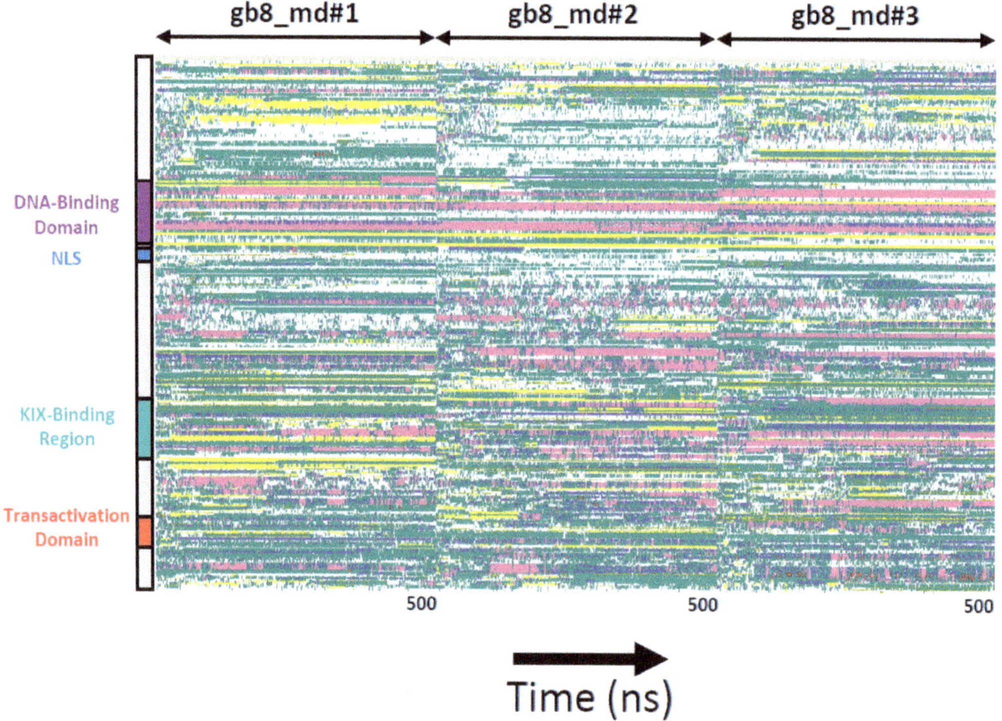

Figure 3. Distribution of secondary structure elements in molecular dynamics (MD) simulations. The distribution of secondary structure elements is shown for the first three MD simulations as representative examples (gb8_md#1, gb8_md#2, gb8_md#3). The vertical axis represents the primary amino acid positions, highlighting functionally relevant domains as

colored boxes). The horizontal axis represents simulation time (1–500 nanoseconds (ns) for each of the three independent simulations). The data are represented in concatenated format to allow direct comparison between the different trajectories. The color codes represent the secondary structures formed in each simulation frame at each position of the primary amino acid sequence (pink, α-helix; dark blue, π-helix; yellow, extended conformation; cyan, turn; white, coil). Data visualized with VMD [34].

Apart from local variation in secondary structures, it is also possible to discern a characteristic higher order of structural organization: both N- and C-terminal portions of FOXO3 are arranged as a mixture of extended and partially folded domains. Projection and superimposition of ten snapshots of the FOXO3/DNA complex at 500 nanosecond (ns) intervals (using the structurally stable DNA-binding domain to align all coordinates) reveals that the two regions adjacent to the DNA-binding domain stretch out and maintain a highly extended structure throughout most of the simulation time (Figure 2B,D; Video S1). I will refer to the flexible linkers located N- and C-terminally to the DNA-binding domain as FL#1 and FL#2, respectively.

The unique structural properties of these linkers will be demonstrated using FL#2 as a specific example (Figure 4). This linker is organized in a bipartite manner with its N-terminal portion (FOXO3$^{260-321}$; FL#2-A) and the C-terminal part (FOXO3$^{322-344}$; FL#2-B) folding in different arrangements. FL#2-A remains in some simulations associated with the NLS and DNA and displays partial helical secondary structure ('FL#2-A compact'; Figure 4A). In other snapshots, FL#2-A is extended and emanates from the DNA-binding domain/NLS without contacts with DNA ('FL#2-A extended'; Figure 4B). FL#2-B also exists in alternative conformations: extended or associated with the more C-terminally located KIX-Binding Region. Accordingly, the length of FL#2 varies between 40 and 70 Å, with an average end-to-end distance of around 55 Å.

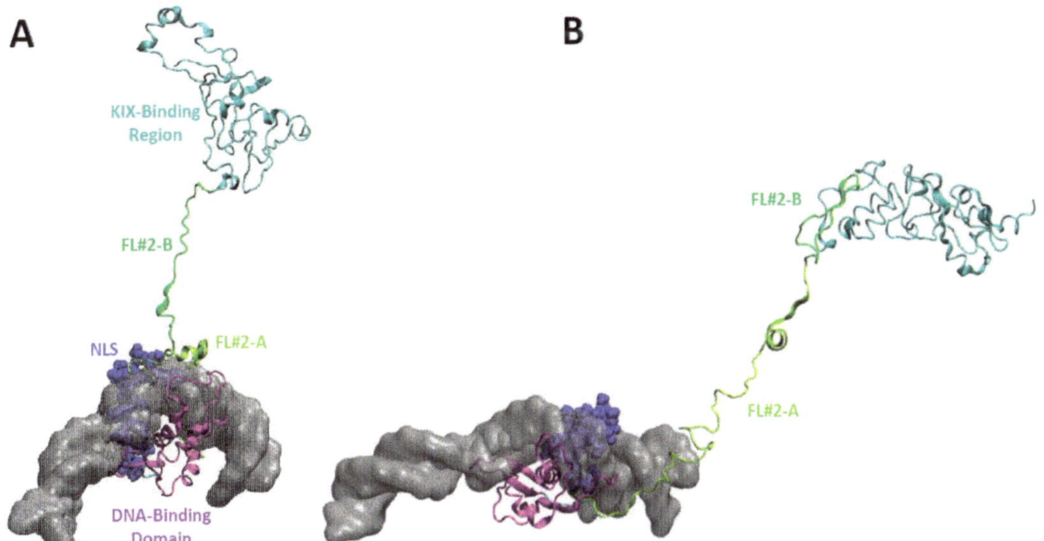

Figure 4. Stretched linker regions emanating from the DNA-binding and nuclear localization sequence (NLS). The DNA-binding domain is shown in a purple cartoon structure bound to DNA represented as a semitransparent surface. The NLS is shown in dark blue as van der Waals spheres. The first part of Flexible Linker #2 (FOXO3$^{260-321}$; FL#2-A) is shown as a lime-green cartoon structure, the second part (FOXO3$^{322-344}$; FL#2-B) in green. The KIX-Binding Region (FOXO3$^{433-508}$) is represented as a cyan cartoon structure. (**A**) FL#2-A folds partially in a helical conformation and remains mostly associated with DNA (FL#2-A compact conformation). FL#2-B is highly extended. (**B**) FL#2-A is present in an extended conformation (FL#2-A extended conformation) and FL#2-B is mostly folded onto the KIX-Binding Region.

The variability in the extension of FL#2 can be visualized effectively in a 'local compaction plot' (LCP). LCPs provide qualitative and quantitative insights into this phenomenon based on the complete trajectory data available. Creation of an LCP is conceptually straightforward: the intramolecular distances between residues spaced apart by a fixed span on the primary amino acids (a 'window') are measured in all trajectory snapshots and plotted. In the examples shown below, a window size of 75 amino acids was chosen to provide a clear visual distinction between folded and unfolded regions. The distance values obtained between the two residues at the beginning and end positions for each window in every frame are plotted along the x-axis. Subsequently, the window is moved by one residue and the pairwise distance measurements are repeated as described above. The 'sliding window' approach is repeated until the end of the sequence represented in the MD simulation is reached. While this creates a massive amount of data (with (tens of) thousands of frames typically present in a MD trajectory and hundreds of sliding window positions), the output can be visualized in a graphical format that is easily interpretable by the human eye by plotting each data point transparently. This allows all measurements to be represented according to their frequency. Frequent measurements are represented as a darker image, whereas alternative conformations occurring at lesser frequency are still shown albeit as a lighter representation. This way of presenting local conformational data is of special value for studying the conformational diversity of MD simulations of IDPs. In the case of FOXO3, the LCP method identifies distinctly the stably folded DNA-binding domain (FOXO3$^{157-237}$) from the remainder of the protein (Figure 5).

A

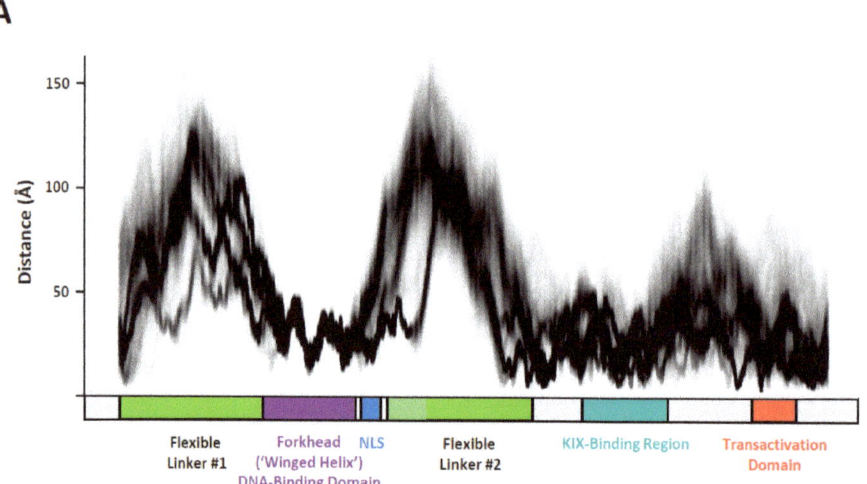

Figure 5. *Cont.*

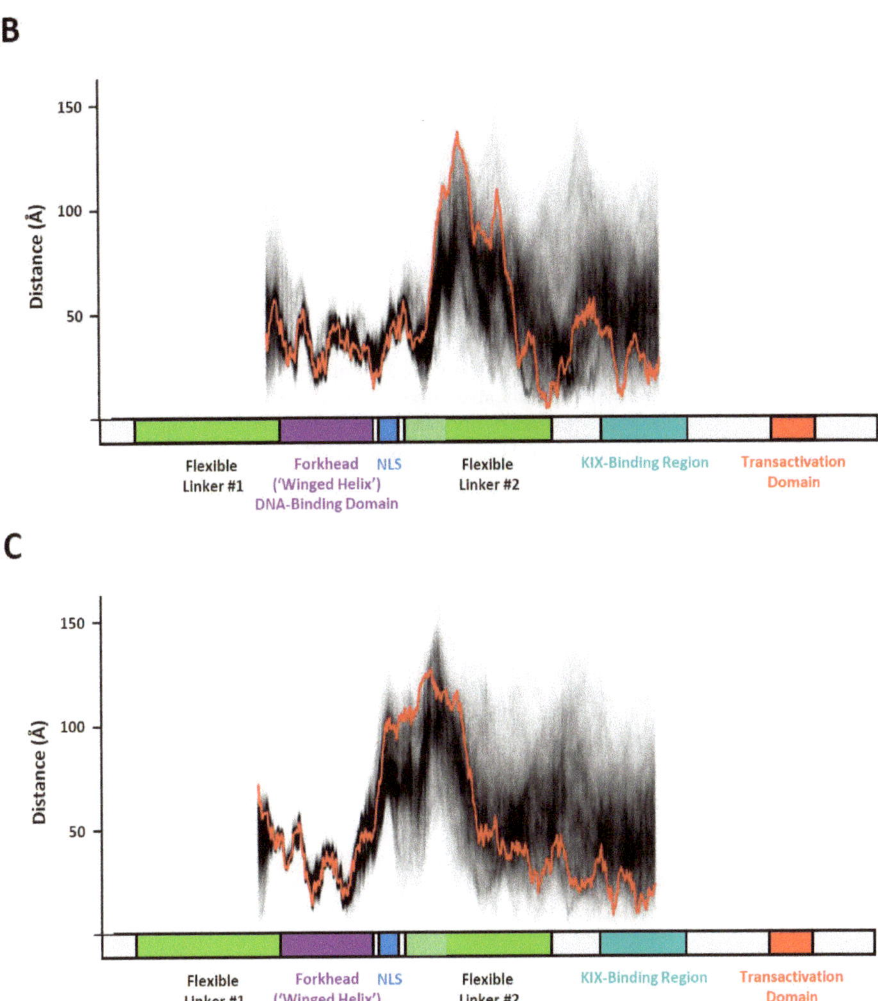

Figure 5. Local compaction plots (LCPs) of FOXO3 molecular dynamics (MD) simulation data. (**A**) LCP analysis of the FOXO3 MD simulation data. The sliding window size for the intramolecular distance measurements is fixed at 75 amino acids. The distance (in Å) between the residues at position 1 and 75 of the window is determined using cpptraj (AMBER simulation package; [32]) for all snapshots of the trajectories. The window is then moved to a new position shifted by one residue. The FOXO3 DNA-binding domain (residues 157 to 237, shown as a purple box) is stably folded and therefore all distance measurements within (and immediately adjacent to it) are quite constant, short (<50 Å) and superimposable on each other to form a dense black line. In contrast, the 'Flexible Linkers' (FL) are distinctly recognizable in the LCP because of the large distances (50–150 Å) identified in the sliding 75 amino acid windows and their more diffuse appearance due to the presence of alternative conformations. (**B**) LCP of the aMD simulation data of FOXO3$^{120-530}$ using the final frame of gb8_md#1 containing FL#2-A in a compact conformation. The LCP trace of the starting structure is shown in red and the traces of the aMD simulations reflecting 100 ns in black. (**C**) LCP of the aMD simulation data of FOXO3 120–530 using the final frame of gb8_md#1 containing FL#2-A in an extended conformation. The LCP trace of the starting structure is shown in red and the traces of the aMD simulations reflecting 100 ns in black.

The FOXO3 DBD (FOXO$^{157-237}$) is highly folded, and therefore all distance measurements within (and immediately adjacent to it) are invariant, short (<50 Å) and superimposable on each other to form a densely black line. In contrast, the 'Flexible Linkers' (FL-1 and FL-2) are distinctly recognizable in the LCP because of the large distances (50–150 Å) identified in the sliding 75 amino acid windows and their more diffuse appearance due to conformational heterogeneity. As pointed out earlier, for FL#2 the simulations reveal at least two distinct patterns, that either include or exclude the NLS sequence located immediately C-terminal to the DBD; whereas in some simulations this structure is still stably structured, in other simulations the NLS becomes part of a more extended FL2 (Figures 4 and 5A). Interestingly, the KIX-Binding Region and C-terminal region containing the Transactivation Domain (TAD) show evidence of notable folding (distances mostly <50–60 Å). The folding is, however, heterogeneous (multiple preferred conformations and containing a subset of unfolded structures as revealed by the diffuse grey lines). This heterogeneity shows that they are still intrinsically disordered, although to a lesser extent than the more flexibly structured linkers FL#1 and FL#2. This example illustrates the simplicity of the concept underlying LCP, while highlighting its uses in interpreting large, complex data sets of MD simulations of IDPs.

Many IDPs contain unfolded polypeptide chains, but these tend to bend and fold into more compact shapes ('random coils') without displaying a comparable degree of rigidity. It was therefore of special interest to look more closely at the structural properties of the rigid linker regions FL-1 and FL-2 present in FOXO3 to determine the structural basis of their enhanced propensity to form extended conformations. Inspection of the amino acid sequence (for example, FL#2B; FOXO3$^{322-344}$, Figure 2C) revealed no unusual features other than an excess of negatively charged amino acid residues. Simulation of FL#2B on its own with identical implicit solvation conditions applied to the intact DNA-FOXO3 complex revealed no distinct rigidity; the distribution of the end-to-end distance of FL#2B was statistically indistinguishable from highly flexible polypeptides of the same length consisting of poly-glycine (poly-G) or alternating glycine and serine residues (poly-G/S) (Figure 2C). It is therefore evident that the extended nature of FL#2B is not encoded directly by its primary amino acid sequence. If we consider the FL regions of FOXO3 to be highly flexible, other physical forces must be responsible for maintaining their extended conformations observed in the simulations of the intact FOXO3/DNA complex. Of all the non-covalent forces determining molecular conformations, only electrostatic forces are capable of long-distance action. A Poisson-Boltzmann surface area (PBSA) plot indeed reveals a significant net negative charge covering the DNA and FOXO3 bound to it (Figure 6). The linkers share this excess negative charge property, thus providing a physicochemical explanation for the extended conformations of FOXO3.

3.2. Confirmation of Electrostatic Repulsion Effect in Explicit Solvent Models

The portion of the molecule comprising the DNA, DBD and initial portion of the C-terminal domain represents one of the clearest examples of the electrostatic repulsion effect (Figure 5A). The use of implicit solvation models is generally reliable and has previously provided good results for other intrinsically disordered domains (such as c-MYC; [23]), but they may display a tendency to generate structures that are folded more compactly than expected. An artefactually high compaction could result in a denser clustering of charges and therefore could be responsible for the electrostatic repulsion effect observed in the implicit solvation simulations of FOXO3. I therefore set up another series of simulations with explicitly solvated models using the TIP4P-D water model optimized specifically for the simulation of IDPs [18]. A portion of FOXO3 lacking the N-terminal IDR (FOXO3$^{120-530}$), but including the DNA-binding domain, the nuclear localization sequence, FL#2 and the more densely folded KIX Binding Region (Figure 1A) was embedded in a TIP4P-D water box and subjected to ten independent accelerated molecular dynamics (aMD) simulations, each lasting 100 nanoseconds. Two sets of simulations, using the final structures of either gb8_md#1 or gb8_md#2—containing FL#2-A in either a compact or extended conformation, respectively, were tested. aMD simulations are thought to represent molecular motions occurring at a time scale that is two to three orders of magnitude higher than the actual simulation time, so the conformational changes observed most likely cover at least one microsecond [30]. TIP4P-D is a water model that was developed specifically for simulating IDPs and produces molecular models that confirm closely to experimentally observed biophysical measurements, such as radius of gyration [18,20,23]. Analysis of the aMD results with the LCP method in an identical manner to the one employed previously for the data from the implicit MD trajectories reveals a comparable outcome (Figure 5B,C). As observed previously, the DBD and NLS remain distinctly folded, and FL-2 extends to values ~~ranging typically~~ in the 80–150 Å range and is thus comparable to the results represented by the starting model (gb8_md#1 and gb8_md#2 after 500 ns, respectively). In the aMD simulations, the region containing the KIX-Binding Region is folded less compactly than with the implicit simulations (as indicated by the larger intramolecular distance measurements in that region; Figure 5B,C), which is in line with the expectations from simulations run the TIP4P-D water model. It can be concluded that the electrostatic repulsion model identified in the implicit simulations is compatible with results from explicitly solvated simulation regimes and therefore reflects a genuine property of the molecular system studied.

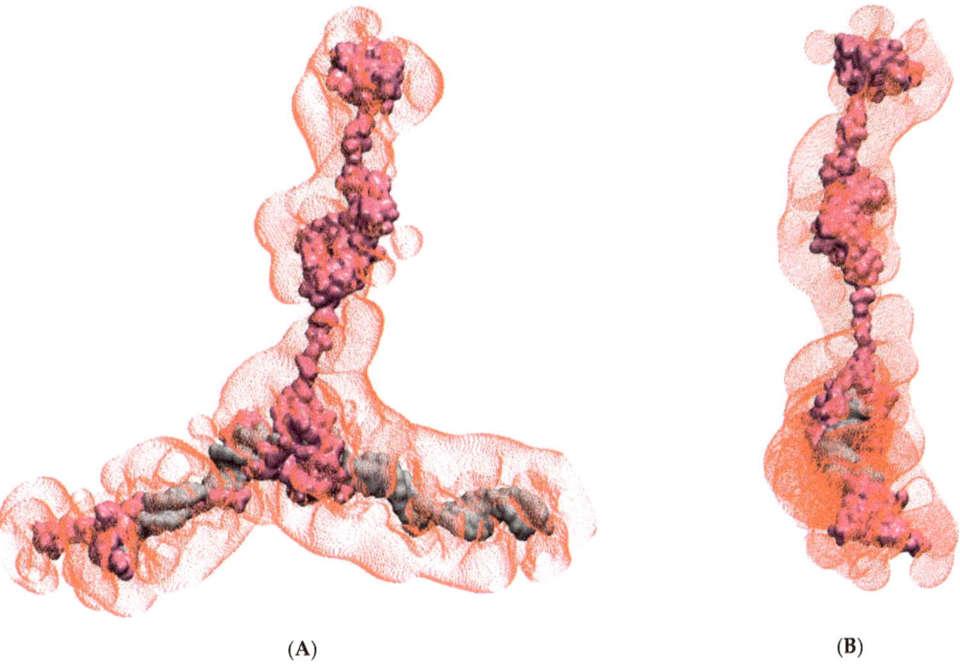

(A)   (B)

Figure 6. Electrostatic surface of FOXO3 bound to DNA. (**A**) Frontal view of the isoelectric surface representation of the FOXO3-DNA complex. The structure of the FOXO3-DNA complex is shown in surface representation (DNA silver, FOXO3 purple). The red dotted surface represents the isosurface with an isovalue of 0.02. (**B**) Same as in (**A**) but rotated by 90° to allow viewing of the complex along the DNA axis.

3.3. Conformational Variability in the KIX-Binding Region and Transactivation Domain

The activation function of FOXO3 depends on two distinct motifs found in the C-terminal IDR, the KIX-Binding Region (FOXO$^{433-508}$) and Transactivation Domain ('TAD'; FOXO$^{606-644}$) (Figure 1A). Although named distinctly, both regions appear to carry out their transactivation functions in a similar manner involving the (simultaneous) binding of their α-helices (FOXO$^{467-478}$ and FOXO$^{622-634}$, respectively) within surface grooves of the KIX domain in the CBP/p300 coactivator. This is presumably achieved by the flexible linker that separates the two motifs within the FOXO3 primary sequence looping out from the coactivator surface [17]. The simulation data presented here are in full agreement with this model. Analysis of the secondary structures formed in the KIX-Binding Region in FOXO3—in the absence of coactivator binding—shows that both interaction motifs are predisposed to form helical conformations, including α-, π- and 3$_{10}$ helices (Figure 7). Especially the centrally located hydrophobic residues known to play a key structural and functional role in these interactions (L^{470}, L^{473}, L^{474} and I^{627}, I^{628}) are in regions with particularly high helical propensity. Such structural predispositions of intrinsically disordered activation domain motifs to form helical structures in the absence of their binding partner is a well-known property of IDPs ('templated folding'; [35,36]) and has been detected in other well-studied model systems of TADs [37,38]. Evidence for a flexible linker connecting the KIX-Binding Region and TAD is also evident from the LCP (Figure 5A).

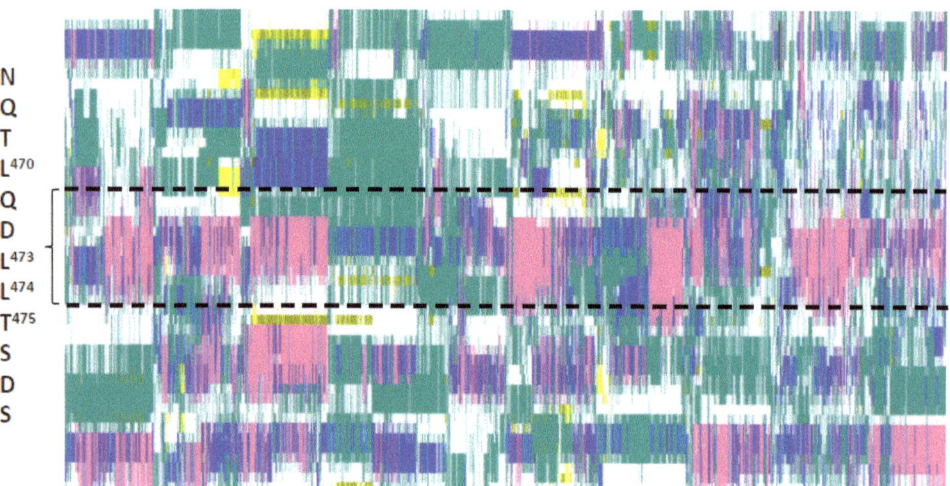

A

Figure 7. *Cont.*

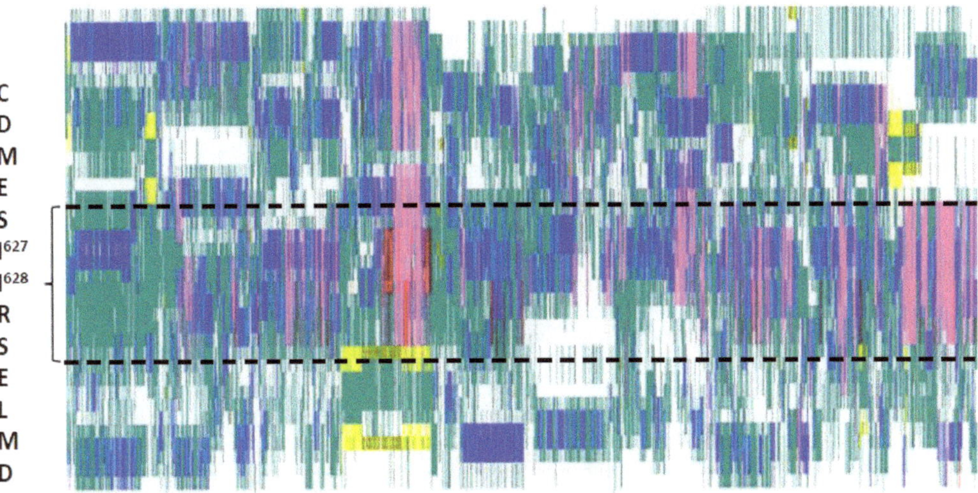

Figure 7. Secondary structures of the KIX-Binding Region and Transactivation Domain (TAD) formed in ten implicit solvation simulations (gb8_md#1—gb8_md#10), each representing 500 nanoseconds. The positions and sequences of α-helical regions shown are based on experimentally determined locations of KIX-binding motifs [17]. (**A**) The region (residues 451 to 500) surrounding the α-helix participating in binding to the KIX-domain of the CBP/p300 coactivator complex is shown. The position and sequence of the interaction helix (FOXO$^{467-478}$) binding to the KIX domain directly is highlighted. (**B**) The region (residues 606 to 644) surrounding another α-helix participating in binding to the KIX-domain of the CBP/p300 coactivator complex is shown. The position and sequence of the interaction helix (FOXO$^{622-634}$) binding to the KIX domain directly is highlighted. The color codes represent the secondary structures formed in each simulation frame at each position of the primary amino acid sequence (pink, α-helix; dark blue, π-helix; red, 3$_{10}$ helix; yellow, extended conformation; cyan, turn; white, coil). Data visualized with VMD [34].

4. Discussion

FOXO3 is an intriguing protein. The high degree of evolutionary conservation of the primary amino acid sequences, especially among vertebrates (Figure S1), is reminiscent of the level of conservation found in structured proteins, where the identities of individual amino acids determine secondary and tertiary structure elements much more rigidly than they do in IDPs. Bioinformatic prediction tools suggest, however, that FOXO3 is mostly disordered outside the centrally located DNA-binding motif (MobiDB [15]; Figure 1A). FOXO3 is therefore either a highly unusual IDP that does not conform to the rules apparently emerging from many other IDPs or contains a higher proportion of sequence-determined structural elements than is evident from the sequence-based bioinformatic analyses. To address this question, a series of fully atomistic MD simulations of a complete FOXO3-DNA complex is presented here. The results confirm the extensive intrinsically disordered nature of the protein outside the DBD as demonstrated by the highly variable and fluctuating secondary structure elements in the N- and C-terminal IDRs (Figures 3 and 5). Interestingly, a systematic analysis of the higher-order structural arrangement shows substantial variation in local folding densities. The DBD is flanked on both sides by highly extended linkers, FL#1 and FL#2, and the more compactly folded KIX-Binding Region and TAD are also separated by a linker of variable length (Figure 2B,D, Figure 5A–C; Video S1). These linkers lack any intrinsic 'stiffness' as determined by their primary sequences, but the domains they are separating are negatively charged and thus repel each other and stretch the connecting unstructured polypeptide sequences (Figure 6). This excess negative charge includes the surface of the DBD that contains a high proportion of positively charged amino acids (Figure 1C), but these are employed primarily for electrostatic binding to DNA. The extended overall structure of the IDRs fully exposes the known functional domains, especially the KIX-Binding Region and TAD near the C-terminus, in a highly flexible manner to protein–protein interactions with the CBP/p300 coactivator complex [17] by providing a large capture radius ('fly casting' mechanism [39]; Video S1). Additionally, the flexibility of FOXO3 may facilitate the subsequent formation of phase separated transcription complexes in conjunction with other components of the transcriptional machinery [40,41]. FOXO3 is a target for several post-translational modifications, especially phosphorylation. Specific phosphorylation at 20 different positions of FOXO3 (mostly in the C-terminal IDR) by various kinases has been documented. Current functional explanations for these phosphorylation events focus on their roles in subcellular localization and regulating transcription activity [42–44]. The electrostatic repulsion mechanism controlling the spatial extension of FOXO3 proposed above suggests that at least some of these phosphorylations may have (additional) conformational effects be controlling intramolecular distances between different parts of FOXO3.

5. Conclusions

The computational simulation of IDPs, such as gene-specific transcription factors, is still technically challenging. While the models obtained may not be perfectly accurate in all aspects, it is clear that some of the most fundamental aspects of FOXO3—the variable extent of folding of the different intrinsically disordered domains and their mutual electrostatic repulsion—is a very robust, yet previously unrecognized property that is revealed reliably through MD simulations.

Supplementary Materials: The following are available online at https://www.mdpi.com/article/10.3390/biom11060856/s1, Figure S1: Sequence alignment of vertebrate and invertebrate orthologs of FOX03, Figure S2: Starting model of FOXO3 bound to DNA, Figure S3: Gaussian kernel density estimation of the phase space determined by root mean square (RMSD) and solvent accessible surface area measurements from the ten independent implicit simulations over 500 ns simulation time. Figure S4: Frequency of secondary structure elements across ten independent molecular dynamics simulations (gb8_md#1—gb8_md#10, 500 nanoseconds each), Video S1: FOXO3-DNA_complex.

Funding: This research received no external funding.

Data Availability Statement: The complete molecular dynamics simulation data and/or copies of the analysis Jupyter notebooks are available from the author on request.

Conflicts of Interest: The author declares no conflict of interest.

References

1. Eijkelenboom, A.P.; Burgering, B.M. FOXOs: Signalling integrators for homeostasis maintenance. *Nat. Rev. Mol. Cell Biol.* **2013**, *14*, 83–97. [CrossRef] [PubMed]
2. Hornsveld, M.; Dansen, T.; Derksen, P.; Burgering, B. Re-evaluating the role of FOXOs in cancer. *Semin. Cancer Biol.* **2018**, *50*, 90–100. [CrossRef] [PubMed]
3. Calissi, G.; Lam, E.W.-F.; Link, W. Therapeutic strategies targeting FOXO transcription factors. *Nat. Rev. Drug Discov.* **2021**, *20*, 21–38. [CrossRef]
4. Yadav, R.K.; Chauhan, A.S.; Zhuang, L.; Gan, B. FoxO transcription factors in cancer metabolism. *Semin. Cancer Biol.* **2018**, *50*, 65–76. [CrossRef]
5. Coomans de Brachene, A.; Demoulin, J.B. FOXO transcription factors in cancer development and therapy. *Cell. Mol. Life Sci.* **2016**, *73*, 1159–1172. [CrossRef] [PubMed]
6. Van Der Heide, L.P.; Hoekman, M.F.; Smidt, M.P. The ins and outs of FoxO shuttling: Mechanisms of FoxO translocation and transcriptional regulation. *Biochem. J.* **2004**, *380*, 297–309. [CrossRef] [PubMed]
7. Brunet, A.; Bonni, A.; Zigmond, M.J.; Lin, M.Z.; Juo, P.; Hu, L.S.; Anderson, M.J.; Arden, K.C.; Blenis, J.; Greenberg, E.M. Akt Promotes Cell Survival by Phosphorylating and Inhibiting a Forkhead Transcription Factor. *Cell* **1999**, *96*, 857–868. [CrossRef]
8. Obsil, T.; Obšilová, V. Structure/function relationships underlying regulation of FOXO transcription factors. *Oncogene* **2008**, *27*, 2263–2275. [CrossRef]
9. Tsai, K.L.; Sun, Y.J.; Huang, C.Y.; Yang, J.Y.; Hung, M.C.; Hsiao, C.D. Crystal structure of the human FOXO3a-DBD/DNA complex suggests the effects of post-translational modification. *Nucleic Acids Res.* **2007**, *35*, 6984–6994. [CrossRef]

10. Benayoun, B.A.; Caburet, S.; Veitia, R.A. Forkhead transcription factors: Key players in health and disease. *Trends Genet.* **2011**, *27*, 224–232. [CrossRef]

11. Ho, K.K.; Myatt, S.S.; Lam, E.W.-F. Many forks in the path: Cycling with FoxO. *Oncogene* **2008**, *27*, 2300–2311. [CrossRef]

12. Liu, Y.; Ao, X.; Ding, W.; Ponnusamy, M.; Wu, W.; Hao, X.; Yu, W.; Wang, Y.; Li, P.; Wang, J. Critical role of FOXO3a in carcinogenesis. *Mol. Cancer* **2018**, *17*, 1–12. [CrossRef]

13. Wang, F.; Marshall, C.B.; Ikura, M. Forkhead followed by disordered tail: The intrinsically disordered regions of FOXO3a. *Intrinsically Disord. Proteins* **2015**, *3*, e1056906. [CrossRef] [PubMed]

14. Dosztányi, Z. Prediction of protein disorder based on IUPred. *Protein Sci.* **2018**, *27*, 331–340. [CrossRef]

15. Piovesan, D.; Necci, M.; Escobedo, N.; Monzon, A.M.; Hatos, A.; Mičetić, I.; Quaglia, F.; Paladin, L.; Ramasamy, P.; Dosztányi, Z.; et al. MobiDB: Intrinsically disordered proteins in 2021. *Nucleic Acids Res.* **2021**, *49*, D361–D367. [CrossRef]

16. So, C.W.; Cleary, M.L. MLL-AFX Requires the Transcriptional Effector Domains of AFX To Transform Myeloid Progenitors and Transdominantly Interfere with Forkhead Protein Function. *Mol. Cell. Biol.* **2002**, *22*, 6542–6552. [CrossRef] [PubMed]

17. Wang, F.; Marshall, C.B.; Yamamoto, K.; Li, G.-Y.; Gasmi-Seabrook, G.M.C.; Okada, H.; Mak, T.W.; Ikura, M. Structures of KIX domain of CBP in complex with two FOXO3a transactivation domains reveal promiscuity and plasticity in coactivator recruitment. *Proc. Natl. Acad. Sci. USA* **2012**, *109*, 6078–6083. [CrossRef] [PubMed]

18. Piana, S.; Donchev, A.G.; Robustelli, P.; Shaw, D.E. Water Dispersion Interactions Strongly Influence Simulated Structural Properties of Disordered Protein States. *J. Phys. Chem. B* **2015**, *119*, 5113–5123. [CrossRef]

19. Mercadante, D.; Milles, S.; Fuertes, G.; Svergun, D.I.; Lemke, E.A.; Gräter, F. Kirkwood–Buff Approach Rescues Overcollapse of a Disordered Protein in Canonical Protein Force Fields. *J. Phys. Chem. B* **2015**, *119*, 7975–7984. [CrossRef] [PubMed]

20. Henriques, J.; Skepö, M. Molecular Dynamics Simulations of Intrinsically Disordered Proteins: On the Accuracy of the TIP4P-D Water Model and the Representativeness of Protein Disorder Models. *J. Chem. Theory Comput.* **2016**, *12*, 3407–3415. [CrossRef] [PubMed]

21. Bhattacharya, S.; Lin, X. Recent Advances in Computational Protocols Addressing Intrinsically Disordered Proteins. *Biomolecules* **2019**, *9*, 146. [CrossRef]

22. Andresen, C.; Helander, S.; Lemak, A.; Farès, C.; Csizmok, V.; Carlsson, J.; Penn, L.Z.; Forman-Kay, J.D.; Arrowsmith, C.; Lundström, P.; et al. Transient structure and dynamics in the disordered c-Myc transactivation domain affect Bin1 binding. *Nucleic Acids Res.* **2012**, *40*, 6353–6366. [CrossRef] [PubMed]

23. Sullivan, S.S.; Weinzierl, R.O.J. Optimization of Molecular Dynamics Simulations of c-MYC(1-88)-An Intrinsically Disordered System. *Life* **2020**, *10*, 109. [CrossRef] [PubMed]

24. Cock, P.J.A.; Antao, T.; Chang, J.T.; Chapman, B.A.; Cox, C.J.; Dalke, A.; Friedberg, I.; Hamelryck, T.; Kauff, F.; Wilczynski, B.; et al. Biopython: Freely available Python tools for computational molecular biology and bioinformatics. *Bioinformatics* **2009**, *25*, 1422–1423. [CrossRef] [PubMed]

25. Boomsma, W.; Frellsen, J.; Harder, T.; Bottaro, S.; Johansson, K.E.; Tian, P.; Stovgaard, K.; Andreetta, C.; Olsson, S.; Valentin, J.B.; et al. PHAISTOS: A framework for Markov chain Monte Carlo simulation and inference of protein structure. *J. Comput. Chem.* **2013**, *34*, 1697–1705. [CrossRef] [PubMed]

26. Salomon-Ferrer, R.; Case, D.A.; Walker, R.C. An overview of the Amber biomolecular simulation package. *Wiley Interdiscip. Rev. Comput. Mol. Sci.* **2013**, *3*, 198–210. [CrossRef]

27. Case, D.A.; Aktulga, H.M.; Belfon, K.; Ben-Shalom, I.Y.; Brozell, S.R.; Cerutti, D.S.; Cheatham, T.E., III; Cruzeiro, V.W.D.; Darden, T.A.; Duke, R.E. *Amber 2020*; University of California: San Francisco, CA, USA, 2020.

28. Maier, J.A.; Martinez, C.; Kasavajhala, K.; Wickstrom, L.; Hauser, K.E.; Simmerling, C. ff14SB: Improving the Accuracy of Protein Side Chain and Backbone Parameters from ff99SB. *J. Chem. Theory Comput.* **2015**, *11*, 3696–3713. [CrossRef] [PubMed]

29. Götz, A.W.; Williamson, M.J.; Xu, D.; Poole, D.; Le Grand, S.; Walker, R.C. Routine Microsecond Molecular Dynamics Simulations with AMBER on GPUs. 1. Generalized Born. *J. Chem. Theory Comput.* **2012**, *8*, 1542–1555. [CrossRef] [PubMed]

30. Pierce, L.C.; Salomon-Ferrer, R.; De Oliveira, C.A.F.; McCammon, J.A.; Walker, R.C. Routine Access to Millisecond Time Scale Events with Accelerated Molecular Dynamics. *J. Chem. Theory Comput.* **2012**, *8*, 2997–3002. [CrossRef] [PubMed]

31. Humphrey, W.; Dalke, A.; Schulten, K. VMD: Visual molecular dynamics. *J. Mol. Graph.* **1996**, *14*, 33–38. [CrossRef]

32. Roe, D.R.; Cheatham, T.E. PTRAJ and CPPTRAJ: Software for Processing and Analysis of Molecular Dynamics Trajectory Data. *J. Chem. Theory Comput.* **2013**, *9*, 3084–3095. [CrossRef]

33. Kabsch, W.; Sander, C. Dictionary of protein secondary structure: Pattern recognition of hydrogen-bonded and geometrical features. *Biopolymers* **1983**, *22*, 2577–2637. [CrossRef] [PubMed]

34. Frishman, D.; Argos, P. Knowledge-based protein secondary structure assignment. *Proteins Struct. Funct. Bioinform.* **1995**, *23*, 566–579. [CrossRef] [PubMed]

35. Toto, A.; Malagrinò, F.; Visconti, L.; Troilo, F.; Pagano, L.; Brunori, M.; Jemth, P.; Gianni, S. Templated folding of intrinsically disordered proteins. *J. Biol. Chem.* **2020**, *295*, 6586–6593. [CrossRef]

36. Toto, A.; Camilloni, C.; Giri, R.; Brunori, M.; Vendruscolo, M.; Gianni, S. Molecular Recognition by Templated Folding of an Intrinsically Disordered Protein. *Sci. Rep.* **2016**, *6*, 21994. [CrossRef] [PubMed]

37. Brzovic, P.S.; Heikaus, C.C.; Kisselev, L.; Vernon, R.; Herbig, E.; Pacheco, D.; Warfield, L.; Littlefield, P.; Baker, D.; Klevit, R.E.; et al. The Acidic Transcription Activator Gcn4 Binds the Mediator Subunit Gal11/Med15 Using a Simple Protein Interface Forming a Fuzzy Complex. *Mol. Cell* **2011**, *44*, 942–953. [CrossRef]

38. Scholes, N.S.; Weinzierl, R.O.J. Molecular Dynamics of "Fuzzy" Transcriptional Activator-Coactivator Interactions. *PLoS Comput. Biol.* **2016**, *12*, e1004935. [CrossRef] [PubMed]
39. Shoemaker, B.A.; Portman, J.J.; Wolynes, P.G. Speeding molecular recognition by using the folding funnel: The fly-casting mechanism. *Proc. Natl. Acad. Sci. USA* **2000**, *97*, 8868–8873. [CrossRef]
40. Peng, L.; Li, E.-M.; Xu, L.-Y. From start to end: Phase separation and transcriptional regulation. *Biochim. Biophys. Acta (BBA) Gene Regul. Mech.* **2020**, *1863*, 194641. [CrossRef] [PubMed]
41. Hnisz, D.; Shrinivas, K.; Young, R.A.; Chakraborty, A.K.; Sharp, P.A. A Phase Separation Model for Transcriptional Control. *Cell* **2017**, *169*, 13–23. [CrossRef]
42. Li, Z.; Zhao, J.; Tikhanovich, I.; Kuravi, S.; Helzberg, J.; Dorko, K.; Roberts, B.; Kumer, S.; A Weinman, S. Serine 574 phosphorylation alters transcriptional programming of FOXO3 by selectively enhancing apoptotic gene expression. *Cell Death Differ.* **2015**, *23*, 583–595. [CrossRef] [PubMed]
43. Wang, X.; Hu, S.; Liu, L. Phosphorylation and acetylation modifications of FOXO3a: Independently or synergistically? *Oncol. Lett.* **2017**, *13*, 2867–2872. [CrossRef] [PubMed]
44. Clavel, S.; Siffroi-Fernandez, S.; Coldefy, A.S.; Boulukos, K.; Pisani, D.F.; Dérijard, B. Regulation of the Intracellular Localization of Foxo3a by Stress-Activated Protein Kinase Signaling Pathways in Skeletal Muscle Cells. *Mol. Cell. Biol.* **2009**, *30*, 470–480. [CrossRef] [PubMed]

Article

Protein–Protein Connections—Oligomer, Amyloid and Protein Complex—By Wide Line ^{1}H NMR

Mónika Bokor and Ágnes Tantos

1   Wigner Research Centre for Physics, Institute for Solid State Physics and Optics, 1121 Budapest, Hungary
2   Research Centre for Natural Sciences, Institute of Enzymology, 1117 Budapest, Hungary; tantos.agnes@ttk.hu
*   Correspondence: bokor.monika@wigner.hu; Tel.: +36-209939420

Abstract: The amount of bonds between constituting parts of a protein aggregate were determined in wild type (WT) and A53T α-synuclein (αS) oligomers, amyloids and in the complex of thymosin-β$_4$–cytoplasmic domain of stabilin-2 (Tβ$_4$-stabilin CTD). A53T αS aggregates have more extensive βsheet contents reflected by constant regions at low potential barriers in difference (to monomers) melting diagrams (*MD*s). Energies of the intermolecular interactions and of secondary structures bonds, formed during polymerization, fall into the 5.41 kJ mol^{-1} ≤ E_a ≤ 5.77 kJ mol^{-1} range for αS aggregates. Monomers lose more mobile hydration water while forming amyloids than oligomers. Part of the strong mobile hydration water–protein bonds break off and these bonding sites of the protein form intermolecular bonds in the aggregates. The new bonds connect the constituting proteins into aggregates. Amyloid–oligomer difference *MD* showed an overall more homogeneous solvent accessible surface of A53T αS amyloids. From the comparison of the nominal sum of the *MD*s of the constituting proteins to the measured *MD* of the Tβ$_4$-stabilin CTD complex, the number of intermolecular bonds connecting constituent proteins into complex is 20(1) H$_2$O/complex. The energies of these bonds are in the 5.40(3) kJ mol^{-1} ≤ E_a ≤ 5.70(5) kJ mol^{-1} range.

Keywords: α-helix; amyloid; β-sheet; hydration; intrinsically disordered proteins; oligomer; protein–protein interactions; wide-line ^{1}H NMR

Citation: Mónika Bokor and Ágnes Tantos Protein–Protein Connections—Oligomer, Amyloid and Protein Complex—By Wide Line ^{1}H NMR. *Biomolecules* **2021**, *11*, 757. https://doi.org/10.3390/biom11050757

Academic Editors: Simona Maria Monti, Giuseppina De Simone and Emma Langella

Received: 17 March 2021
Accepted: 14 May 2021
Published: 18 May 2021

Publisher's Note: MDPI stays neutral with regard to jurisdictional claims in published maps and institutional affiliations.

1. Introduction

One of the main reasons intrinsically disordered proteins (IDPs) are so important in different physiological processes is that they are capable of forming a bewildering variety of interactions. They can undergo induced folding or unfolding, they are able to form physiological or pathological aggregates, and even form fuzzy complexes, where disorder is retained during the interaction [1]. Describing the molecular details of the different interaction types of IPDs is crucial for the clear understanding their mechanisms of function and for the interference with the pathological processes caused by their erroneous behavior.

In our work presented here, two IDP interaction systems were studied to gain information about the bonds holding the protein associations—oligomer, amyloid, and complex—together. One system contains wild type and mutant α-synuclein (αS) in the forms of oligomers and amyloids. Detailed structural and preparation information can be found in [2,3]. The other system consists of two proteins—thymosin-β$_4$ (Tβ$_4$) and cytoplasmic domain of stabilin-2 (stabilin CTD)—which form a 1:1 complex (see [4,5] for structural information and for details of preparation).

αS is a 140-amino-acid protein mainly located at presynaptic terminals and is abundant in the brain [6]. The exact physiological role of αS remains unclear, but it's thought to be involved in the regulation of synaptic vesicle mobility [7] through binding to VAMP [8]. Pathological aggregation of αS is the leading cause of Parkinson's disease (PD), and some mutations of αS, such as A53T, cause familial occurrences of the disease [9,10]. A53T mutation results in faster αS accumulation [11]. When the balance between the production and clearance of αS is disturbed, the monomeric αS aggregates and misfolds into oligomers, then amyloid fibrils and finally Lewy bodies [12]. The exact pathogenesis of PD is still unknown, but more and more evidence suggests that the oligomeric form of αS plays a vital part in the pathogenesis of PD [12–16]. Historically, αS has been considered as an IDP in the healthy, physiological state [17,18]. IDPs exist as dynamic protein ensembles with fluxional structures under physiological conditions that can adopt different conformations depending on their external stimuli, thus enabling the complex signaling and regulatory requirements of higher biological systems [19,20].

The structural traits that lie behind the stronger amyloidogenic propensity of the disease-related mutants are intensively studied and have revealed many important features. The helical content of wild type (WT) αS is minimally affected by the A53T mutation, while there is an increase in the β-sheet content of the A53T mutant-type in comparison to the WT protein [18,21–24]. Long-range intramolecular interactions between the different regions either are less abundant or disappear upon A53T mutation [25–27], indicating that the NAC (non-AB component) region is more solvent-exposed upon A53T mutation [28]. As this segment is indispensable for the amyloid formation, its easier accessibility explains the increased aggregation propensity of the A53T mutant [22,23,26,29–33]. The structures of the A53T αS are thermodynamically more preferred than the structures of the WT αS [26].

Our understanding of amyloid structure lately has been improved a lot by innovations in cryo-electron microscopy, electron diffraction and solid-state NMR. The results show expected cross-β amyloid structure and an unexpectedly diverse and complex amyloid fold [34]. The amyloid cores are constituted of extensive mainchain hydrogen bonding between adjacent β-strands within the stacked layers, and close interdigitation of sidechains. The amyloid fibrils are dynamic, with monomers and/or oligomers dissociating from their ends [35,36].

Isolated Tβ₄, stabilin CTD and their 1:1 complex create a system where the bonds formed in the complex can be studied.

Tβ₄ is an IDP with moonlighting functions. It can sequester actin [37] and it has important role in the regulation of the formation and modulation of the actin cytoskeleton [38]. Many different functions were attributed to Tβ₄. Such functions are, e.g., endothelial cell differentiation [39], angiogenesis [40] and wound repair [41]. Tβ₄ can increase the concentration of metalloproteinases, which is important in cell migration [41], the underlying process of which is still unclear. The multiple functions of Tβ₄ are probably related to that, it is an IDP and has very little transient secondary structure in solution phase [42]. It binds often weakly to its physiological partners, and forms structurally heterogeneous complexes with them [43], but it can fold upon binding.

Stabilin-2 is an endocytic receptor for hyaluronic acid, and its C-terminal domain (CTD) binds Tβ₄ [4,44]. The observation that Tβ₄ colocalizes with stabilin-2 in the phagocytic cup suggests that the complex of Tβ₄ with stabilin-2 CTD is involved in the phagocytosis of apoptotic cells [45]. It was also shown that knockdown or overexpression of Tβ₄ decreased and enhanced the phagocytic activity of stabilin-2, respectively. While the exact molecular mechanism of this is yet unclear, it was found that Tβ₄ engages in multiple fuzzy interactions with stabilin-2 CTD. It is capable of mediating its distinct yet specific interactions without adapting distinct folded structures in the different complexes [42]. Weak binding was confirmed between Tβ₄ and stabilin-2 CTD and it was suggested that the proteins become slightly more disordered in the complex [4,5].

The purpose of this experimental study is to determine the amount and quality of bonds between constituting parts of a protein aggregate in wild type (WT) and A53T α-synuclein (αS) oligomers, amyloids and in the complex of thymosin-β₄–cytoplasmic domain of stabilin-2. The selection of the studied proteins represents a system in which the constituents are different in their degree of polymerization and another system where the constituents interact with each other. This work presents as novelty compared to the previous works [2–5] the differences between α-Ss of different degrees of polymerization and sheds light on the bonds between the constituting proteins in the Tβ_4-stabilin CTD complex.

2. Materials and Methods

The studied proteins were prepared, and their qualities were checked as in [2–5]. Briefly, untagged versions of α-synuclein variants were expressed in *E. coli* cells and purified using anion exchange chromatography, while stabilin CTD was expressed with an N-terminal His-tag and purified on a Ni-NTA affinity column. The wide-line ^{1}H NMR signals (see Appendix A and Supporting Information of [3]) on the time scale are composed of more components of different origins (ice, protein, mobile water) and time constants. The intensities of the components with the longest decaying-time constants give the amounts of the mobile hydration water. Melting diagrams (*MDs*) are the relative amounts of the proteins' mobile water measured by intensities of wide-line ^{1}H NMR signals vs. normalized functional temperature. The mobile hydration differences were calculated from *MDs* of WT and A53T αS monomers, oligomers and amyloids in [3], and from *MDs* of Tβ_4, stabilin-2 CTD and their 1:1 complex in [4,5]. In this work, potential barriers corresponding to normalized functional temperature are used. Potential barriers are calculated from functional (or absolute) temperature by first normalizing it with the melting point of water (273.15 K) and then scaling this normalized temperature with the melting heat of ice ($Q = 6.01$ kJ mol^{-1}), $E_a = T \cdot Q / T_m$, where T is the absolute temperature and T_m is the melting point of water. The relative amount of mobile water is converted to molar amount of mobile water per amino acid residues, *naa*. The *MDs* are given as *naa* vs. E_a. The motion of water is controlled by the potential barrier.

3. Results

3.1. The α-Synuclein System

Mobile hydration differences (monomer *MD* subtracted from oligomer (o-m) and amyloid (a-m) *MDs*) were calculated for αS variants of different polymerization degrees. The difference *MDs* for the o-m and the *a-m* cases are very similar for each variant (Figure 1).

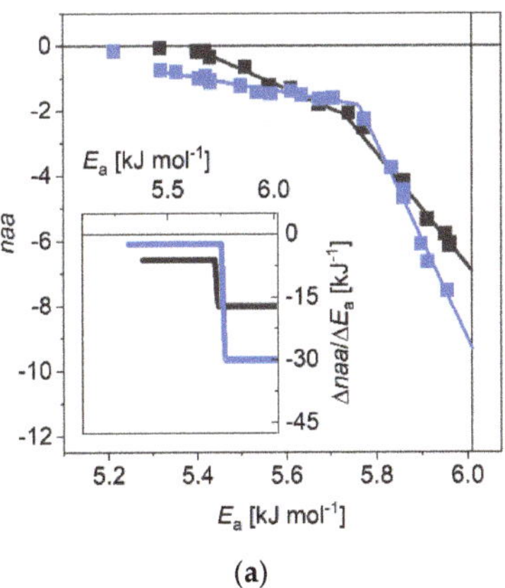
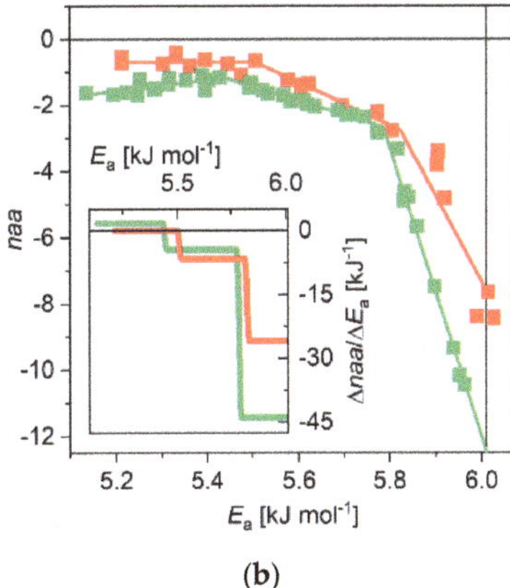

(a)　　　　　　　　　　**(b)**

Figure 1. (a) Wild type α-synuclein, oligomer-monomer (grey) and amyloid–monomer (blue) melting diagrams; (b) A53T α-synuclein, oligomer–monomer (red) and amyloid–monomer (green) melting diagrams. Mol H$_2$O per mol amino acid residue (*naa*) vs. potential barrier. Lines are guides to the eye. The insets are the derivative forms of the difference melting diagrams. Mol H$_2$O per mol amino acid residue (*naa*) vs. potential barrier. Lines are guides to the eye. The insets are the derivative forms of the difference melting diagrams.

The derivative ratios are depicted in the insets of the figures and they reflect very sensitively the changes in the trends of the melting curves. However, they have characteristically different shapes for the WT and the A53T variants. While the WT αS o-m difference *MD* has two sections with distinct slopes, A53T αS has three. The additional low-potential-barrier (E_a) sections in A53T difference curves can be attributed to the excess β-sheet contents [22–25] of the mutant compared to WT. The additional β-sheet content is reflected by the low E_a constant *naa* (mol H_2O per mol amino acid residue) or only little raising *naa* sections in o-m and a-m differences, respectively (Figure 1b). This is supported by the inverse preferences to form helices for alanine and to support β-sheet structures for threonine [46], i.e., A53T αS is more prone to have β-sheets than WT αS. This suggestion is affirmed by the comparison of o-m and a-m differences for the WT and the A53T αS variants (Figure 2).

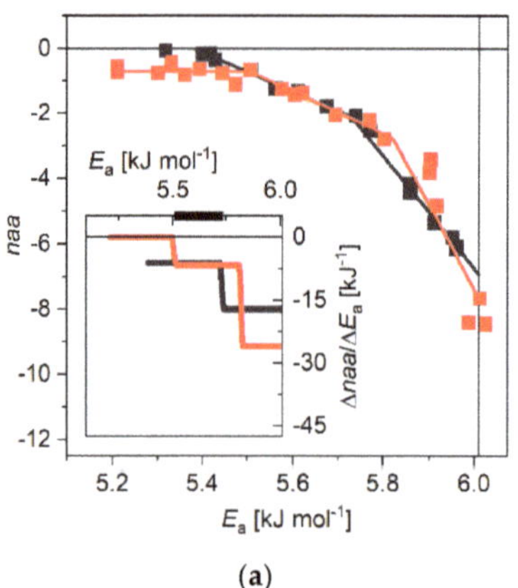
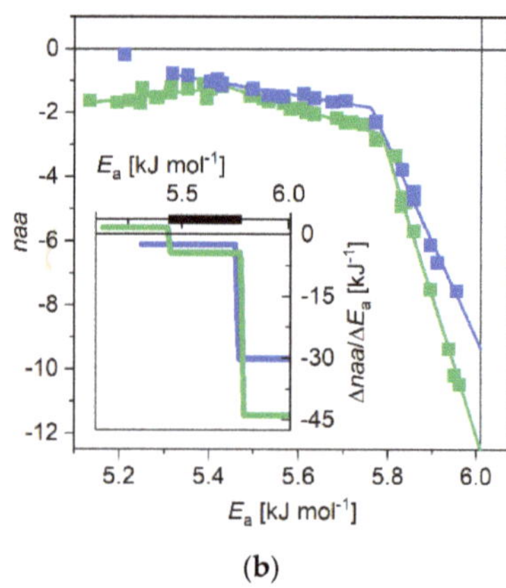

Figure 2. (a) Oligomer-monomer melting diagrams for wild type (grey) and A53T (red) α-synuclein; (b) Amyloid-monomer melting diagrams for wild type (blue) and A53T (green) α-synuclein. Mol H_2O per mol amino acid residue (*naa*) vs. potential barrier. Lines are guides to the eye. The insets are the derivative forms of the difference melting diagrams. The thick black lines denote the common sections on the derivative melting diagrams.

The WT and the A53T o-m differences are equal in the potential barrier region 5.41 kJ mol^{-1} ≤ E_a ≤ 5.77 kJ mol^{-1} (Figure 2a). This region belongs to interactions incorporating the monomers into oligomers or amyloids and to the secondary structures forming in the process of polymerization. The loss of mobile hydration water is the most intensive at potential barrier values near to the melting point of bulk water (6.01 kJ mol^{-1}). The slope of the decrease is less steep in the case of o-m curves than for o-a curves (Figures 1 and 2). In this region, close to the melting point of bulk water, the most strongly bound water molecules start to move. The mobile hydration water loss is the greatest for the a-m curves, here is where we can detect the largest difference compared to the monomers. The monomers lose much more hydration water, which is visible as mobile for NMR, while forming amyloids than in the formation of oligomers. Some of the strong mobile hydration water–protein bonds are disrupted, allowing the protein to form intermolecular connections in the oligomers and amyloids. These new bonds are the ones that hold the oligomers and amyloids together. In the ~5.8 kJ mol^{-1} to 6.01 kJ mol^{-1} potential barrier region, A53T αS amyloids lose the most mobile hydration water, Δnaa = 9.8. While the WT αS amyloids lose less mobile hydration water, Δnaa = 7.5. The oligomers make approximately half as much water–protein bonds in their formation, namely Δnaa = 4.9 in the A53T variants and it is Δnaa = 4.0 in the WT αS oligomers. This suggests a more compact form of A53T oligomers, partially explaining why this mutant forms amyloids at a much faster pace than the wild type αS.

The (amyloid–oligomer) mobile hydration difference shows that oligomers are more hydrated than amyloids (Figure 3). As mentioned earlier, this is directly connected to the differences in the compactness of the two protein structures. The difference varies between *naa* = −0.9 and *naa* = −1.8, except at 5.85 kJ mol^{-1} ≤ E_a ≤ 6.01 kJ mol^{-1}, where it extends to as large values as *naa* $\approx$ −13. The difference changes more for WT than for A53T αS, the average is −1.40(5) with 0.08 variance for WT αS in the 4.96 kJ mol^{-1} $\leq E_a \leq$ 5.85 kJ mol^{-1} potential barrier range while the average is −1.27(4) with 0.04 variance for A53T αS in the same energy range. This means an overall more homogeneous solvent accessible surface of A53T αS. It can be the result of the amyloids having smaller solvent accessible surface per monomer units compared to oligomers.

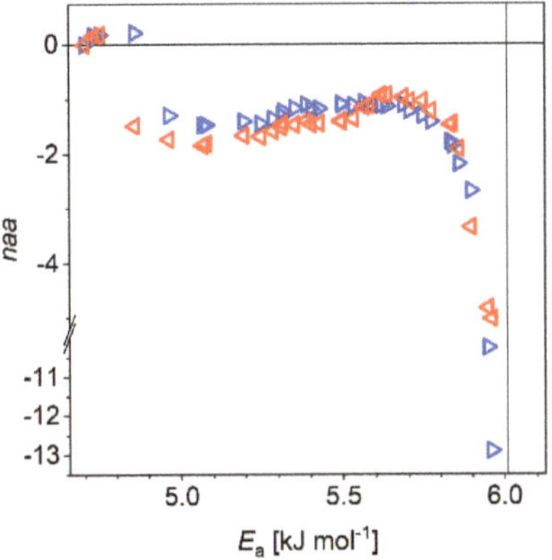

Figure 3. Wild type (red triangles) and A53T (blue triangles) α-synuclein amyloid-oligomer difference melting diagrams. Mol H_2O per mol amino acid residue (*naa*) vs. potential barrier.

3.2. The Stabilin CTD-Tβ_4 System

The number of the intermolecular bonds connecting Tβ_4 and stabilin CTD was determined, to build 1:1 protein complex. The mobile hydration values per amino acid residues were simply added together for the constituting proteins at each potential barrier values, i.e., the *MDs* of these proteins were numerically added. The sum was compared to the measured *MD* of the 1:1 protein complex (Figure 4). The comparison revealed that the sum is greater than the measured values of the 1:1 complex between 5.40(3) kJ mol^{-1} $\leq E_a \leq$ 5.70(5) kJ mol^{-1} potential barrier values. Outside of this range, the sum and measured *MDs* coincide. The difference equals to *naa* = 0.21(1) or 20(1) H_2O/complex and gives the number of mobile hydration water to protein bonds broken by the formation of the 1:1 protein complex. These broken bonds mean interaction sites that then make new intramolecular bonds, which hold the 1:1 protein complex together. The energy of the bonds falls in the potential-barrier interval of the broken bonds.

Given the sizes of the two proteins, the number of intermolecular bonds is low. Since both of the partners are disordered in isolation, this few intramolecular bonds are not capable of restricting the movement of the two proteins, resulting in a fuzzy complex. In this setup, both Tβ_4 and stabilin CTD retain most of their structural disorder even in the bound state. Fuzzy complexes, where some level of flexibility is observed after binding, are frequently found in nature [47]. In most cases, when an IDP makes connections with a globular protein, fuzziness is limited to a certain level, but binding of two IDPs can lead to the formation of complexes with extreme fuzziness—where the partners remain mostly disordered [5]. These types of interactions pose a challenge for structural characterization, as most methods are unable to pick up the subtle changes that occur upon binding in such a way. Our results clearly show that Tβ_4 and stabilin CTD are capable of interacting with each other and remain almost completely disordered in their complex.

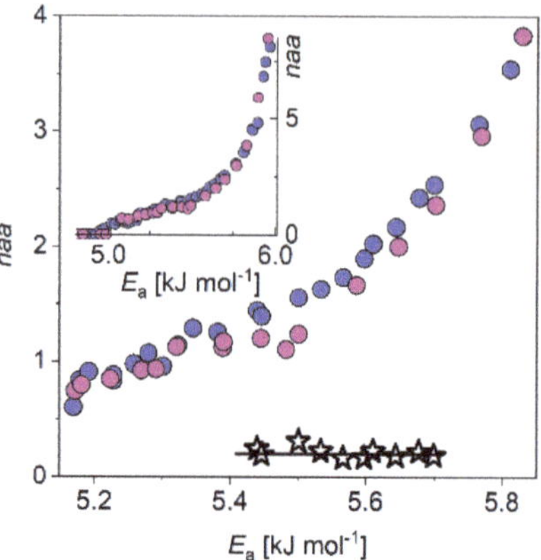

Figure 4. Sum of the measured melting diagrams of isolated thymosin-β_4 and stabilin-2 CTD (blue circles). Measured melting diagram of thymosin-β_4-stabilin-2 CTD complex (magenta circles). Their difference is given by black stars and line. Inset: whole potential barrier range from the no mobile hydration water point to the melting point of water.

4. Discussion

The A53T mutants have surplus β-sheets to the wild type α synucleins according to FTIR, ss-NMR and single-molecule force spectroscopy experiments [21,23,24]. In our results, the additional β-sheet content is indicated by the extra low-potential-barrier constant or almost constant *MD* difference section (Figure 2). The bonds between the monomers in oligomers or amyloids and the secondary structures specific to the aggregates appear as coinciding sections of WT and A53T αSs (thick black lines in the insets of Figure 2). The most strongly bound water molecules start to move close to the melting point of bulk water. Oligomers lose more mobile hydration water than the amyloids as it can be seen on the (amyloid–oligomer) hydration difference. On place of the lost protein–water bonds, new intermolecular connections form during aggregation, which bond the monomers into oligomers or amyloids. The A53T oligomers are more compact than the WT ones according to the (amyloid–oligomer) *MD* differences (Figure 3). This compactness partially gives reason the mutant to form amyloids faster. The oligomers are more hydrated as a result. The A53T αS has more homogeneous solvent accessible surface per monomer units compared to WT αS because, e.g., the mobile hydration in the (amyloid–oligomer) *MD* differences are smaller with half as big variance. The (amyloid–oligomer) *MD* differences being negative, reflect the amyloids having smaller solvent accessible surface per monomer units compared to oligomers.

The sum of the *MD*s of the proteins Tβ$_4$ and stabilin CTD is greater than the measured *MD* of the 1:1 protein complex by *naa* = 0.21(1) or 20(1) H$_2$O/complex (Figure 4). The Tβ$_4$ and stabilin CTD interact with each other while remain almost completely disordered in their complex. Binding of two IDPs like Tβ$_4$ and stabilin CTD leads to the formation of a complex with extreme fuzziness [5], the partner proteins remain disordered in the complex. Both constituting proteins in the 1:1 complex retain most of their structural disorder. Tβ$_4$ and stabilin CTD connect to each other with only a small number of intermolecular bonds. The intermolecular bonds are not enough numerous to limit the movement of the two proteins in the complex.

5. Conclusions

Mobile hydration differences (monomer *MD* subtracted from oligomer and amyloid *MD*s) were calculated for αS variants of different polymerization degrees. The derivative ratios of the *MD*s reflect very sensitively the changes in the trends of the melting curves, and they have characteristically different shapes for the WT and the A53T variants. The constant low E_a

sections in A53T αS difference curves for oligomers and amyloids are caused by the excess β-sheet contents of the mutant compared to WT. The 5.41 kJ mol^{-1} $\leq E_a \leq$ 5.77 kJ mol^{-1} difference *MD* regions belong to interactions linking the monomers into oligomers or amyloids and to the bonds of secondary structures forming in the process of polymerization. The monomers lose much more mobile hydration water while forming amyloids than in the formation of oligomers, reflecting fundamental structural differences between oligomers and amyloids. A part of the strong mobile hydration water–protein bonds are lost and these bonding sites of the protein form intermolecular bonds in the oligomers and amyloids. The new bonds being created connect the constituting proteins into oligomers or amyloids. The amyloid–oligomer difference *MD* showed that there is an overall more homogeneous solvent accessible surface of A53T αS amyloids than the WT variants, suggesting a more densely packed amyloid structure.

We gained information on the intermolecular bonds constructing the αS oligomer and amyloid aggregates from the monomers.

The number of the intermolecular bonds was determined, which connect Tβ$_4$ and stabilin CTD in the thymosin-β$_4$–stabilin-2 CTD complex. It was found that there are *naa* = 0.21(1) or 20(1) H$_2$O/complex such bonds and their bonding energy falls into the 5.40 kJ mol^{-1} $\leq E_a \leq$ 5.70 kJ mol^{-1} range. These results were gained from the comparison of the nominal sum of the *MDs* of the constituting proteins to the measured *MD* of the thymosin-β$_4$–stabilin-2 CTD complex and provide further insight into the molecular background of their fuzzy complexes.

Author Contributions: Conceptualization, M.B. and Á.T.; methodology, M.B.; validation, M.B. and Á.T.; formal analysis, M.B.; investigation, M.B.; data curation, M.B.; writing—original draft preparation, M.B. and Á.T.; writing—review and editing, M.B. and Á.T.; visualization, M.B.; supervision, M.B.; funding acquisition, Á.T. All authors have read and agreed to the published version of the manuscript.

Funding: This research was funded by National Research, Development and Innovation Office, Hungary grant number K-125340 (Á.T.). Wigner Research Centre for Physics covered publication costs.

Institutional Review Board Statement: Not applicable.

Informed Consent Statement: Not applicable.

Acknowledgments: Kaálmaán Tompa is gratefully thanked for his support.

Conflicts of Interest: The authors declare no conflict of interest.

Appendix A

Wide-Line ^{1}H NMR-Spectrometry

Frozen protein solutions are measured with wide-line ^{1}H NMR to get melting diagrams in the temperature range from −70 °C to +20 °C. The free induction decay (FID) signal can be separated into differently decaying components. The slowly decaying component is generated by the ^{1}H nuclei of the mobile hydration water. The extrapolated amplitude to time zero of each FID-component is proportional to the amount of each ^{1}H fraction.

The mobile water molecules are found at the surface of the protein molecules. The melting diagram gives the amount of mobile water measured by NMR as a function of temperature. The corresponding potential barrier values to temperature, E_a are given as the normalized fundamental temperature multiplied with the molar heat of fusion for water ($E_a = T/273.15$ K $\times$ 6.01 kJ mol^{-1}, normalized fundamental temperature is multiplied with the molar heat of fusion of water [48]), thus providing an energy scale.

The amount of mobile, i.e., molten water is measured directly as a fraction of the total water content of the protein solution, *n* vs. *T*. It is converted into the number of mobile water molecules per amino acid residue, *naa*. The *naa* vs. E_a data can provide quantitative analysis beyond spectroscopy.

The melting curve can be formally described as a series expansion (Supporting Information of [5]),

$$n\left(T_{fn}\right) \;=\; A + B\left(T_{fn} - T_{fn1}\right) + C\left(T_{fn} - T_{fn2}\right)^2 + D\left(T_{fn} - T_{fn3}\right)^3 + \ldots$$

The parameter T_{fn1} gives the temperature where the thermal trend of the *MD* switches between constant and linearly increasing. Likewise, the trend changes from linear to quadratic at T_{fn2} and it becomes cubic at T_{fn3}.

The fitting procedure is carried out in sections. It starts with the lowest temperatures. If there is a constant n region, then the values of A are determined and kept for further analysis. Then, follows the linear section and parameters B and T_{fn1} are gained and used in sections of higher powers. The next section is increasing quadratically and C and T_{fn2} is fitted, then the cubic section is next with D and T_{fn3}.

References

1. Tompa, P.; Schad, E.; Tantos, A.; Kalmar, L. Intrinsically disordered proteins: Emerging interaction specialists. *Curr. Opin. Struct. Biol.* **2015**, *35*, 49–59. [CrossRef]
2. Hazy, E.; Bokor, M.; Kalmar, L.; Gelencser, A.; Kamasa, P.; Han, K.-H.; Tompa, K.; Tompa, P. Distinct Hydration Properties of Wild-Type and Familial Point Mutant A53T of α-Synuclein Associated with Parkinson's Disease. *Biophys. J.* **2011**, *101*, 2260–2266. [CrossRef] [PubMed]
3. Bokor, M.; Tantos, Á.; Tompa, P.; Han, K.-H.; Tompa, K. WT and A53T α-Synuclein Systems: Melting Diagram and Its New Interpretation. *Int. J. Mol. Sci.* **2020**, *21*, 3997. [CrossRef] [PubMed]
4. Tantos, A.; Szabó, B.; Láng, A.; Varga, Z.; Tsylonok, M.; Bokor, M.; Verebelyi, T.; Kamasa, P.; Tompa, K.; Perczel, A.; et al. Multiple fuzzy interactions in the moonlighting function of thymosin-β4. *Intrinsically Disord. Proteins* **2013**, *1*, e26204. [CrossRef] [PubMed]
5. Bokor, M.; Tantos, Á.; Mészáros, A.; Jenei, B.; Haminda, R.; Tompa, P.; Tompa, K. Molecular Motions and Interactions in Aqueous Solutions of Thymosin-β 4, Stabilin CTD and Their 1: 1 Complex, Studied by 1 H-NMR Spectroscopy. *ChemPhysChem* **2020**, *21*, 1420–1428. [CrossRef] [PubMed]
6. Goedert, M. Alpha-synuclein and neurodegenerative diseases. *Nat. Rev. Neurosci.* **2001**, *2*, 492–501. [CrossRef]
7. Scott, D.; Roy, S. -Synuclein Inhibits Intersynaptic Vesicle Mobility and Maintains Recycling-Pool Homeostasis. *J. Neurosci.* **2012**, *32*, 10129–10135. [CrossRef]
8. Sun, J.; Wang, L.; Bao, H.; Premi, S.; Das, U.; Chapman, E.R.; Roy, S. Functional cooperation of α-synuclein and VAMP2 in synaptic vesicle recycling. *Proc. Natl. Acad. Sci. USA* **2019**, *116*, 11113–11115. [CrossRef]
9. Jo, E.; Fuller, N.; Rand, R.; George-Hyslop, P.S.; Fraser, P.E. Defective membrane interactions of familial Parkinson's disease mutant A30P α-synuclein. *J. Mol. Biol.* **2002**, *315*, 799–807. [CrossRef]
10. Spira, P.J.; Sharpe, D.M.; Halliday, G.; Cavanagh, J.; Nicholson, G.A. Clinical and pathological features of a parkinsonian syndrome in a family with an Ala53Thr ?-synuclein mutation. *Ann. Neurol.* **2001**, *49*, 313–319. [CrossRef]
11. Marchalonis, J. Lymphocyte surface immunoglobulins. *Science* **1975**, *190*, 20–29. [CrossRef]
12. Burré, J.; Sharma, M.; Südhof, T.C. Definition of a Molecular Pathway Mediating -Synuclein Neurotoxicity. *J. Neurosci.* **2015**, *35*, 5221–5232. [CrossRef] [PubMed]
13. Outeiro, T.F.; Putcha, P.; Tetzlaff, J.E.; Spoelgen, R.; Koker, M.; Carvalho, F.; Hyman, B.T.; McLean, P.J. Formation of Toxic Oligomeric α-Synuclein Species in Living Cells. *PLoS ONE* **2008**, *3*, e1867. [CrossRef]
14. Winner, B.; Jappelli, R.; Maji, S.K.; Desplats, P.A.; Boyer, L.; Aigner, S.; Hetzer, C.; Loher, T.; Vilar, M.; Campioni, S.; et al. In vivo demonstration that -synuclein oligomers are toxic. *Proc. Natl. Acad. Sci. USA* **2011**, *108*, 4194–4199. [CrossRef] [PubMed]
15. Cremades, N.; Cohen, S.I.; Deas, E.; Abramov, A.Y.; Chen, A.Y.; Orte, A.; Sandal, M.; Clarke, R.W.; Dunne, P.; Aprile, F.A.; et al. Direct Observation of the Interconversion of Normal and Toxic Forms of α-Synuclein. *Cell* **2012**, *149*, 1048–1059. [CrossRef] [PubMed]
16. Prots, I.; Veber, V.; Brey, S.; Campioni, S.; Buder, K.; Riek, R.; Böhm, K.J.; Winner, B. α-Synuclein Oligomers Impair Neuronal Microtubule-Kinesin Interplay. *J. Biol. Chem.* **2013**, *288*, 21742–21754. [CrossRef]
17. Dunker, A.K.; Silman, I.; Uversky, V.N.; Sussman, J.L. Function and structure of inherently disordered proteins. *Curr. Opin. Struct. Biol.* **2008**, *18*, 756–764. [CrossRef]
18. Mor, D.E.; Ugras, S.E.; Daniels, M.J.; Ischiropoulos, H. Dynamic structural flexibility of α-synuclein. *Neurobiol. Dis.* **2016**, *88*, 66–74. [CrossRef]
19. Minde, D.-P.; Dunker, A.K.; Lilley, K.S. Time, space, and disorder in the expanding proteome universe. *Proteome* **2017**, *17*, 1600399. [CrossRef]
20. Chen, J.; Kriwacki, R.W. Intrinsically Disordered Proteins: Structure, Function and Therapeutics. *J. Mol. Biol.* **2018**, *430*, 2275–2277. [CrossRef]
21. Li, J.; Uversky, V.N.; Fink, A.L. Conformational Behavior of Human α-Synuclein is Modulated by Familial Parkinson's Disease Point Mutations A30P and A53T. *NeuroToxicology* **2002**, *23*, 553–567. [CrossRef]

22. Conway, K.A.; Harper, J.D.; Lansbury, P.T. Accelerated in vitro fibril formation by a mutant α-synuclein linked to early-onset Parkinson disease. *Nat. Med.* **1998**, *4*, 1318–1320. [CrossRef] [PubMed]
23. Heise, H.; Celej, M.S.; Becker, S.; Riedel, D.; Pelah, A.; Kumar, A.; Jovin, T.M.; Baldus, M. Solid-State NMR Reveals Structural Differences between Fibrils of Wild-Type and Disease-Related A53T Mutant α-Synuclein. *J. Mol. Biol.* **2008**, *380*, 444–450. [CrossRef] [PubMed]
24. Brucale, M.; Sandal, M.; Di Maio, S.; Rampioni, A.; Tessari, I.; Tosatto, L.; Bisaglia, M.; Bubacco, L.; Samorì, B. Pathogenic Mutations Shift the Equilibria of α-Synuclein Single Molecules towards Structured Conformers. *ChemBioChem* **2009**, *10*, 176–183. [CrossRef]
25. Giasson, B.I.; Uryu, K.; Trojanowski, J.Q.; Lee, V.M.-Y. Mutant and Wild Type Human α-Synucleins Assemble into Elongated Filaments with Distinct Morphologies in Vitro. *J. Biol. Chem.* **1999**, *274*, 7619–7622. [CrossRef]
26. Bertoncini, C.W.; Jung, Y.-S.; Fernandez, C.O.; Hoyer, W.; Griesinger, C.; Jovin, T.M.; Zweckstetter, M. From the Cover: Release of long-range tertiary interactions potentiates aggregation of natively unstructured -synuclein. *Proc. Natl. Acad. Sci. USA* **2005**, *102*, 1430–1435. [CrossRef]
27. Narhi, L.; Wood, S.J.; Steavenson, S.; Jiang, Y.; Wu, G.M.; Anafi, D.; Kaufman, S.A.; Martin, F.; Sitney, K.; Denis, P.; et al. Both Familial Parkinson's Disease Mutations Accelerate α-Synuclein Aggregation. *J. Biol. Chem.* **1999**, *274*, 9843–9846. [CrossRef]
28. Coskuner, O.; Wise-Scira, O. Structures and Free Energy Landscapes of the A53T Mutant-Type α-Synuclein Protein and Impact of A53T Mutation on the Structures of the Wild-Type α-Synuclein Protein with Dynamics. *ACS Chem. Neurosci.* **2013**, *4*, 1101–1113. [CrossRef] [PubMed]
29. Serpell, L.C.; Berriman, J.; Jakes, R.; Goedert, M.; Crowther, R.A. Fiber diffraction of synthetic alpha -synuclein filaments shows amyloid-like cross-beta conformation. *Proc. Natl. Acad. Sci. USA* **2000**, *97*, 4897–4902. [CrossRef]
30. Kamiyoshihara, T.; Kojima, M.; Ueda, K.; Tashiro, M.; Shimotakahara, S. Observation of multiple intermediates in α-synuclein fibril formation by singular value decomposition analysis. *Biochem. Biophys. Res. Commun.* **2007**, *355*, 398–403. [CrossRef]
31. Lashuel, H.A.; Petre, B.M.; Wall, J.; Simon, M.; Nowak, R.J.; Walz, T.; Lansbury, P.T. α-Synuclein, Especially the Parkinson's Disease-associated Mutants, Forms Pore-like Annular and Tubular Protofibrils. *J. Mol. Biol.* **2002**, *322*, 1089–1102. [CrossRef]
32. Conway, K.A.; Lee, S.-J.; Rochet, J.-C.; Ding, T.T.; Williamson, R.E.; Lansbury, P.T. Acceleration of oligomerization, not fibrillization, is a shared property of both alpha -synuclein mutations linked to early-onset Parkinson's disease: Implications for pathogenesis and therapy. *Proc. Natl. Acad. Sci. USA* **2000**, *97*, 571–576. [CrossRef] [PubMed]
33. Ono, K.; Ikeda, T.; Takasaki, J.-I.; Yamada, M. Familial Parkinson disease mutations influence α-synuclein assembly. *Neurobiol. Dis.* **2011**, *43*, 715–724. [CrossRef]
34. Gallardo, R.; Ranson, N.A.; Radford, S.E. Amyloid structures: Much more than just a cross-β fold. *Curr. Opin. Struct. Biol.* **2020**, *60*, 7–16. [CrossRef] [PubMed]
35. Hoshino, M. Fibril formation from the amyloid-β peptide is governed by a dynamic equilibrium involving association and dissociation of the monomer. *Biophys. Rev.* **2017**, *9*, 9–16. [CrossRef]
36. Tipping, K.W.; Karamanos, T.K.; Jakhria, T.; Iadanza, M.G.; Goodchild, S.C.; Tuma, R.; Ranson, N.A.; Hewitt, E.W.; Radford, S.E. pH-induced molecular shedding drives the formation of amyloid fibril-derived oligomers. *Proc. Natl. Acad. Sci. USA* **2015**, *112*, 5691–5696. [CrossRef] [PubMed]
37. Safer, D.; Elzinga, M.; Nachmias, V.T. Thymosin β4 and Fx, an actin-sequestering peptide, are indistinguishable. *J. Biol. Chem.* **1991**, *266*, 4029–4032. [CrossRef]
38. Yarmola, E.G.; Klimenko, E.S.; Fujita, G.; Bubb, M.R. Thymosin beta4: Actin Regulation and More. *Ann. N. Y. Acad. Sci.* **2007**, *1112*, 76–85. [CrossRef]
39. Grant, D.S.; Rose, W.; Yaen, C.; Goldstein, A.; Martinez, J.; Kleinman, H. Thymosin β4 enhances endothelial cell differentiation and angiogenesis. *Angiogenesis* **1999**, *3*, 125–135. [CrossRef]
40. Smart, N.; Rossdeutsch, A.; Riley, P.R. Thymosin β4 and angiogenesis: Modes of action and therapeutic potential. *Angiogenesis* **2007**, *10*, 229–241. [CrossRef]
41. Qiu, P.; Sosne, G.; Kurpakus-Wheater, M. Matrix metalloproteinase activity is necessary for thymosin beta 4 promotion of epithelial cell migration. *J. Cell. Physiol.* **2007**, *212*, 165–173. [CrossRef] [PubMed]
42. Domanski, M.; Hertzog, M.; Coutant, J.; Gutsche-Perelroizen, I.; Bontems, F.; Carlier, M.-F.; Guittet, E.; van Heijenoort, C. Coupling of Folding and Binding of Thymosin β4 upon Interaction with Monomeric Actin Monitored by Nuclear Magnetic Resonance. *J. Biol. Chem.* **2004**, *279*, 23637–23645. [CrossRef] [PubMed]
43. Hertzog, M.; van Heijenoort, C.; Didry, D.; Gaudier, M.; Coutant, J.; Gigant, B.; Didelot, G.; Préat, T.; Knossow, M.; Guittet, E.; et al. The β-Thymosin/WH2 Domain: Structural Basis for the Switch from Inhibition to Promotion of Actin Assembly. *Cell* **2004**, *117*, 611–623. [CrossRef]
44. Ho, J.H.-C.; Tseng, K.-C.; Ma, W.-H.; Chen, K.-H.; Lee, O.K.-S.; Su, Y. Thymosin beta-4 upregulates anti-oxidative enzymes and protects human cornea epithelial cells against oxidative damage. *Br. J. Ophthalmol.* **2008**, *92*, 992–997. [CrossRef] [PubMed]
45. Lee, S.-J.; So, I.-S.; Park, S.-Y.; Kim, I.-S. Thymosin β4 is involved in stabilin-2-mediated apoptotic cell engulfment. *FEBS Lett.* **2008**, *582*, 2161–2166. [CrossRef] [PubMed]
46. Podoly, E.; Hanin, G.; Soreq, H. Alanine-to-threonine substitutions and amyloid diseases: Butyrylcholinesterase as a case study. *Chem. Interac.* **2010**, *187*, 64–71. [CrossRef] [PubMed]

47. Wang, W.; Wang, D. Extreme Fuzziness: Direct Interactions between Two IDPs. *Biomolecules* **2019**, *9*, 81. [CrossRef] [PubMed]
48. CRC. *Handbook of Chemistry and Physics*, 101st ed.; CRC Press: Boca Raton, FL, USA, 2020; ISBN 9780367417246.

 *biomolecules*

Article

Structural Analysis of the cl-Par-4 Tumor Suppressor as a Function of Ionic Environment

Krishna K. Raut, Komala Ponniah and Steven M. Pascal

Department of Chemistry and Biochemistry, Old Dominion University, Norfolk, VA 23529, USA; kraut@odu.edu (K.K.R.); kponniah@odu.edu (K.P.)
* Correspondence: spascal@odu.edu; Tel.: +1-757-683-6763

Abstract: Prostate apoptosis response-4 (Par-4) is a proapoptotic tumor suppressor protein that has been linked to a large number of cancers. This 38 kilodalton (kDa) protein has been shown to be predominantly intrinsically disordered in vitro. In vivo, Par-4 is cleaved by caspase-3 at Asp-131 to generate the 25 kDa functionally active cleaved Par-4 protein (cl-Par-4) that inhibits NF-κB-mediated cell survival pathways and causes selective apoptosis in tumor cells. Here, we have employed circular dichroism (CD) spectroscopy and dynamic light scattering (DLS) to assess the effects of various monovalent and divalent salts upon the conformation of cl-Par-4 in vitro. We have previously shown that high levels of sodium can induce the cl-Par-4 fragment to form highly compact, highly helical tetramers in vitro. Spectral characteristics suggest that most or at least much of the helical content in these tetramers are non-coiled coils. Here, we have shown that potassium produces a similar effect as was previously reported for sodium and that magnesium salts also produce a similar conformation effect, but at an approximately five times lower ionic concentration. We have also shown that anion identity has far less influence than does cation identity. The degree of helicity induced by each of these salts suggests that the "Selective for Apoptosis in Cancer cells" (SAC) domain—the region of Par-4 that is most indispensable for its apoptotic function—is likely to be helical in cl-Par-4 under the studied high salt conditions. Furthermore, we have shown that under medium-strength ionic conditions, a combination of high molecular weight aggregates and smaller particles form and that the smaller particles are also highly helical, resembling at least in secondary structure, the tetramers found at high salt.

Keywords: intrinsically disordered protein (IDP); prostate apoptosis response-4 (Par-4); tumor suppressor; circular dichroism (CD) spectroscopy; dynamic light scattering (DLS); aggregation; coiled-coil; caspase

Citation: Krishna K. Raut, Komala Ponniah and Steven M. Pascal Structural Analysis of the cl-Par-4 Tumor Suppressor as a Function of Ionic Environment. *Biomolecules* **2021**, *11*, 386. https://doi.org/10.3390/biom11030386

Academic Editor: Simona Maria Monti

Received: 27 January 2021
Accepted: 26 February 2021
Published: 5 March 2021

1. Introduction

Prostate apoptosis response-4 (Par-4) is a tumor suppressor protein that is capable of selectively inducing apoptosis in cancer cells while leaving healthy cells unaffected [1]. Par-4 was first identified through its down-regulation in therapy-resistant prostate cancer cells and has since been shown to be ubiquitously expressed, localized in the cytoplasm, the nucleus and extracellularly, and linked to a variety of cancers [2–5]. The *par-4* gene maps to human chromosome 12q21 and encodes an approximately 40 kilodalton (kDa) protein of 340 amino acids [6]. While Par-4 has been shown to be largely intrinsically disordered in vitro, it does contain several identifiable domains and features. The first identified domain was the C-terminal coiled-coil, which is highly conserved across humans, mice, and rats [7]. This region contains a small nuclear export signal (NES). The coiled-coil is responsible for Par-4 dimerization, and for most known interactions with cellular proteins. Two nuclear localization signals are found in the N-terminal half of the protein, though only the 2nd, NLS2, has been shown to affect localization [8,9]. Par-4 also contains a unique and highly conserved "Selective for Apoptosis in Cancer cells" (SAC) domain, which is the minimal region required to induce apoptosis [8,9].

Although Par-4 is expressed in both normal and cancer cells, only cancer cells are typically susceptible to the killing effect of this protein. The two best-documented factors controlling this selectivity are (i) GRP78 levels and (ii) PKA levels. (i) Extra-cellular Par-4 can enter cancer cells by binding to the surface receptor GRP78 via the Par-4 SAC domain [10]. GRP78 levels in healthy cells are typically very low or even undetectable, abrogating entry of extracellular Par-4 into healthy cells. (ii) Protein Kinase A (PKA) is required to phosphorylate intracellular Par-4 at residue T163, which is within the SAC domain [4,11,12]. This phosphorylation event leads to the inhibition of PKC-ζ by Par-4, which in turn suppresses phosphorylation of FADD by PKC-ζ. Unphosphorylated FADD is required for DISC (Death-Inducing Signal Complex) formation at the plasma membrane, which triggers the caspase-8-mediated apoptotic pathway [9,10]. As with GRP78, the basal level of PKA in healthy cells is typically low and insufficient to cause phosphorylation of Par-4, thereby preventing DISC formation in healthy cells.

Interestingly, one key event downstream of caspase-8 activation is the activation of caspase-3, which in turn cleaves intracellular Par-4. Caspase-3 cleaves Par-4 after D131, producing two fragments: a 15 kDa amino-terminal fragment (called PAF for Par-4 Amino terminal Fragment) and a 25 kDa carboxy-terminal fragment (called cl-Par-4 for cleaved) [8,13]. Assisted by NLS2, cl-Par-4 translocates to the nucleus, where it blocks cell survival pathways by inhibiting transcriptional regulators such as NF-κB [14,15].

Intrinsically disordered proteins (IDPs) or intrinsically disordered protein regions (IDPRs) possess specific features such as a high percentage of charged residues and a low percentage of hydrophobic amino acids [16–18]. These exceptional characteristics lead to a disordered conformation under typical physiological conditions [19]. A number of environmental factors can, however, influence the folding of these proteins including temperature, pH, ionic strength, post-translational modification including both phosphorylation and targeted cleavage by proteases, and interaction with natural ligands or protein partners [16,20,21]. Significant evidence has arisen showing enhanced folding of IDPs with increased ion concentration and ionic charge density, though unfolding can also be triggered by the same factors [22].

We have previously shown that high levels of sodium chloride or acidic pH can induce the cl-Par-4 fragment to form highly compact, highly helical structures in vitro, at high micromolar protein concentrations [23,24]. The ionic conditions required to favor these small, well-folded particles are outside of the range of conditions normally found in cells. This led to the hypothesis that multiple factors, including multiple ion types, may combine forces to influence Par-4 conformation. To begin to test this hypothesis, we here have examined and compared the structural characteristics of cl-Par-4 as a function of sodium chloride, potassium chloride, magnesium chloride, and magnesium sulfate concentration. The results shed light on the role of cations versus anions, their charge densities, and also on the role of the caspase-induced cleavage event that produces cl-Par-4, in influencing Par-4 structure, and subsequently, Par-4 localization and function.

2. Materials and Methods

2.1. Expression and Purification of cl-Par-4

BL21 (DE3) *Escherichia coli* cells transfected with a codon-optimized construct of cl-Par-4 in the modified expression vector H-MBP-3C [25] were grown in Luria-Bertani (LB) media with 100 µg/mL ampicillin at 37 °C until an OD_{600} of 0.8–0.9 was reached. The cells were then induced for protein expression with the addition of 0.5 mM isopropyl thio-β-D-galactoside (IPTG) and grown at 15 °C until OD_{600} reached 1.5–1.6. The harvested cells were lysed by a combination of mechanical (sonication) and enzymatic (lysozyme) methods and extracted protein was purified with immobilized metal affinity chromatography (IMAC) using a HisTrap HP 5 mL column (GE Healthcare, Uppsala, Sweden). The His-MBP tag was cleaved from cl-Par-4 by treating the protein with the His-tagged 3C-protease enzyme. The purified protein was then further dialyzed against the storage buffer (10 mM Tris, 1 M NaCl, 1 mM TCEP, pH 7.0) and concentrated by centrifugation using a 10 kDa MWCO

Vivaspin Turbo 15 (Sartorius, Epsom, UK). The protein concentration was determined by taking absorbance readings at 280 nm using a BioDrop DUO (BioDrop Ltd., Cambridge, UK) and the theoretical extinction coefficient of 6400 $M^{-1}cm^{-1}$. The protein was finally lyophilized using a FreeZone Freeze Dryer (LABCONCO Corporation, Kansas City, MO, USA), stored at $-80\ ^\circ$C, and re-solubilized in sterile distilled water when required. In our experience, in vitro CD and DLS data obtained with cl-Par-4 samples prepared, lyophilized, and redissolved as described above were indistinguishable from data obtained with fresh (never lyophilized) samples.

2.2. Circular Dichroism Spectroscopy

Circular dichroism (CD) spectra of the protein were obtained on a J-815 CD spectrometer (Jasco, Easton, MD, USA). The cl-Par-4 samples were at a concentration of 0.2 mg/mL in Tris-HCl buffer (10 mM Tris, 1 mM DTT, pH 7.0) with varying concentrations of monovalent and divalent cations present (10–1000 mM). Each cl-Par-4 sample used for CD spectroscopy contained 40 mM NaCl as a residual due to the dilution from the storage buffer (10 mM Tris, 1 mM TCEP, and 1 M NaCl, pH 7.0). For the filtered samples, sample preparation was identical except for the filtration of the sample through a 0.45-micron Nalgene SFCA membrane syringe filter (Thermo Fisher Scientific, Rochester, NY, USA) immediately prior to CD spectroscopy. For the centrifuged samples, sample preparation was identical except that samples were centrifuged for 5 min at 10,000 RPM ($9615\times g$) using a Sorvall Legend Micro 21R Centrifuge (Thermo Fisher Scientific, Germany), and the supernatant was used for the CD spectroscopy. CD data were recorded over a 400–190 nm wavelength range at a scan speed of 20 nm/min using a 1 mm pathlength quartz cuvettes at 25 $^\circ$C. A buffer blank was subtracted from the spectra, which were then smoothed using a means-movement function of 25 nm and converted to molar ellipticity The deconvolution of CD spectra for estimation of secondary structure content was done using the SELCON3 algorithm through the DichroWeb server [26].

2.3. Dynamic Light Scattering

Dynamic light scattering (DLS) data were recorded using a NanoBrook Omni particle sizer and zeta potential analyzer (Brookhaven Instruments Corporation, Holtsville, NY, USA). Protein samples were at a concentration of 0.2 mg/mL in Tris buffer (10 mM Tris, 1 mM DTT, pH 7.0) with varying concentrations of monovalent and divalent cations (10–1000 mM). Filtered and centrifuged samples were prepared as described above. The data were recorded at 25 $^\circ$C using a standard diode laser at 640 nm wavelength, scattering angle of 173°, and plastic cuvettes of 1 cm pathlength. Five scans were recorded for each sample and hydrodynamic radii (Stokes' radii) were calculated from the mean effective diameter obtained from the summary statistical report of the NanoBrook software.

2.4. SDS-PAGE

Sodium dodecyl sulfate-polyacrylamide gel electrophoresis (SDS-PAGE) gels of 4–12% were prepared using a gel casting system (Bio-Rad Laboratories, Inc., Irvine, CA, USA). Filtered samples were prepared as mentioned above. The samples, each containing initially 0.2 mg/mL of the protein (filtered samples contained less protein), were prepared for loading by combining 16 μL of each sample with 4 μL of 5× sample loading dye, mixed well, and heated at 90 $^\circ$C for 2 min. Next, 20 μL of each sample was loaded into separate wells in the gel. The gels were run in 1× SDS-PAGE buffer (Tris base, glycine, SDS, and water) using the PowerPac Basic device (Bio-Rad Laboratories, Inc., Irvine, CA, USA) for 20 min with a low voltage of 50 V and then 60 min with a high voltage of 150 V. The gels were then stained for 5 h in Coomassie stain (0.2% R-250, 7.5% acetic acid, 50% ethanol and water) and de-stained overnight in de-staining solution (7.5% acetic acid, 50% ethanol, and water). Visualization of the gel was done on a white lightbox.

3. Results

3.1. Disorder Prediction in cl-Par-4

The cl-Par-4 fragment is the functionally active fragment of the Par-4 protein that enters the nucleus. This region of the protein can be further subdivided as follows: (Figure 1a): the SAC (Selective for Apoptosis induction in Cancer cells) domain, the CC (Coiled-Coil) domain, and the Linker between these two domains. In addition, the SAC domain contains an NLS (Nuclear Localization Signal) near its N-terminus, and the CC domain contains an LZ (Leucine Zipper) domain which comprises its C-terminal half [23]. Prediction of disorder in these domains was done by using DISOPRED3, which utilizes X-ray diffraction databases (Figure 1b) [27,28]. Considering a probability of 0.5 (dashed line) as the demarcation line, the analysis clearly predicts order in the CC domain and disorder in the Linker, while the SAC domain has intermediate character, approaching the order/disorder line. Previous sequence analysis also showed mixed order/disorder propensity and some helical character in the SAC domain, and so order in the SAC domain would not be surprising, under the right conditions [23].

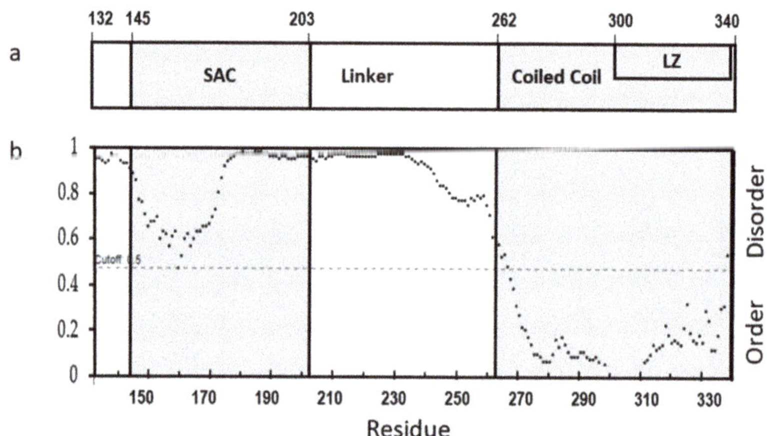

Figure 1. (**a**) Schematic diagram showing domains of cleaved Par-4 protein (cl-Par-4) and (**b**) Disorder prediction in cl-Par-4 using DISOPRED3.

3.2. Differential Effect of Monovalent & Divalent Ions on the Structure of cl-Par-4

We have previously shown that at a high NaCl concentration, cl-Par-4 forms a highly helical tetramer with mostly non-coiled coil helices [24]. Here, we investigated the effects of high levels of a different monovalent cation (potassium), with a different size and charge density. The relative effects of NaCl and KCl on CD spectra of cl-Par-4 are shown in Figure 2a. All cl-Par-4 samples were at a concentration of 0.2 mg/mL in Tris-buffer of pH 7.0 containing various concentrations of either NaCl or KCl.

At a 500 mM salt concentration, CD spectra were characteristic of α-helical content with intense dichroism minima at 222 nm and 208 nm [29,30]. At lower salt concentrations, the intensity of the CD spectra was reduced, which will be discussed further in Section 3.4. At all concentrations, the effect of Na vs. K cations appeared to be similar.

For isolated alpha-helices, an intensity ratio of greater than 1 for the two minima at θ_{222} and θ_{208} indicates coiled-coil formation [24,29,31]. By these criteria, it appears that the 500 mM salt sample did not form a significant coiled-coil, but that the lower salt samples may. An alternative interpretation of conformation at low and medium salt concentrations is presented in the discussion section.

The effects of the divalent cation magnesium on the structural conformation of cl-Par-4 were also investigated via CD spectroscopy. Sample preparation was similar to above, but with various concentrations of $MgSO_4$ instead of NaCl or KCl. CD spectra (Figure 2b) showed a similar trend as obtained with monovalent cations, with higher intensity at higher salt concentration. Again, the θ_{222}: θ_{208} ratio suggests little coiled-coil formation at high salt concentration and the possibility of coiled-coil at low salt concentration.

However, the most striking difference between monovalent and divalent cation salt samples was in the concentrations of demarcation: all samples with 75 mM and higher $MgSO_4$ showed similar, intense, highly helical but non-coiled coil CD spectra, whereas only the 500 mM monovalent salt samples produced such a result. Intermediate spectra occurred for 50 mM $MgSO_4$, while analogous intermediate spectra occurred at a much higher monovalent salt concentration of 250 mM NaCl or KCl. The effects of $MgSO_4$ and KCl are directly compared in Figure 2c. These results suggest that the divalent magnesium cation is five or more times as effective as monovalent cations in influencing the structure of cl-Par-4.

Next, the effect of monovalent versus divalent anions on the structure of cl-Par-4 was investigated by comparing samples prepared with $MgSO_4$ versus $MgCl_2$. The results showed little difference between like concentrations of $MgSO_4$ and $MgCl_2$ (Figure 2d), suggesting that cation identity exerts a far greater influence on cl-Par-4 structure than does anion identity.

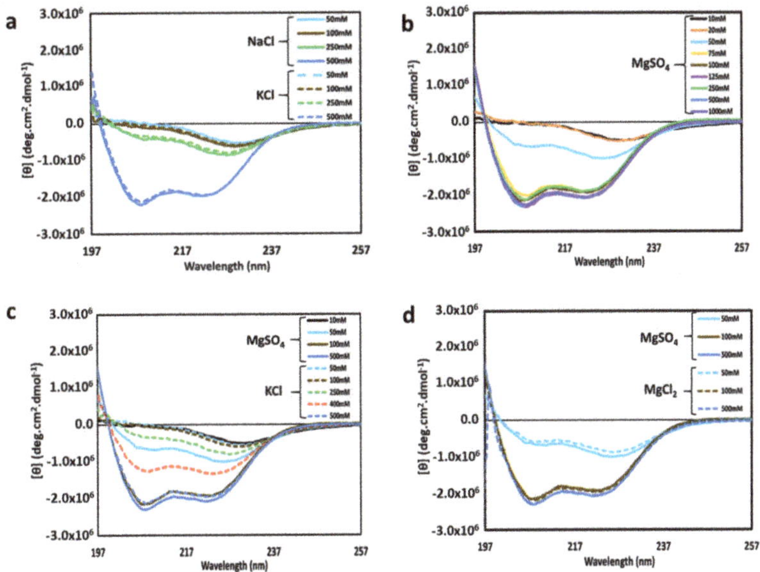

Figure 2. Overlay of circular dichroism (CD) spectra of cl-Par-4 at various concentrations of the following salts: (**a**) NaCl (solid lines) and KCl (dashed lines); (**b**) $MgSO_4$; (**c**) $MgSO_4$ (solid lines) and KCl (dashed lines); (**d**) $MgSO_4$ (solid lines) and $MgCl_2$ (dashed lines). The salt concentrations are indicated by the color legend within each panel. All spectra were recorded at pH 7.0.

3.3. Effect of Monovalent & Divalent Ions on Hydrodynamic Size of cl-Par-4

The size of protein molecules in solution (hydrodynamic size) provides information about conformation. Small size indicates a compact and well-folded conformation whereas large size indicates a highly self-associated, or aggregated conformation, that may or may not also be significantly disordered. The hydrodynamic size distribution of cl-Par-4 was investigated using dynamic light scattering (DLS). Sample preparation was similar to above, but with the following concentrations of NaCl: 50, 100, 250, and 500 mM or $MgSO_4$: 10, 50, 100, and 500 mM. The result showed relatively small particles (Stokes' radius, Rs $\approx$ 200 nm) in the sample with 500 mM of NaCl (Figure 3a). The observed hydrodynamic sizes were significantly larger (Rs $\approx$ 1000 nm) at the three lower NaCl concentrations (50, 100, and

250 mM), indicating the presence of large aggregates. The polydispersity values associated with DLS measurements (see numbers above bars in Figure 3a) correlate strongly with Stokes' radii, indicating that in general, samples with larger particles also contain a broader range of particle sizes.

A similar trend of smaller particle size at higher salt was observed in the presence of $MgSO_4$ (Figure 3b). However, the transition between small (Rs ≈ 200 nm) and large (Rs ≈ 1000 nm) particles apparently occurs somewhere between $MgSO_4$ concentrations of 50 mM and 100 mM, suggesting that the divalent magnesium ion is approximately five times more potent than the monovalent cation in influencing cl-Par-4 particle size. Again, the largest Stokes' radii (at 10 mM $MgSO_4$) were associated with the broadest range of particle sizes, as shown by polydispersity measurements in Figure 3b. However, in general, less polydispersity was seen in the $MgSO_4$ samples than in the NaCl samples, indicating that even when large particles were present, magnesium induced a more uniform particle size than did sodium.

Note that large particle size as detected by DLS (Figure 3a,b) corresponded closely with reduced CD intensity (Figure 2a,b). However, the correspondence was not perfect. In particular, note that DLS detected similar particle sizes at 50, 100, and 250 mM NaCl (Figure 3a), but that 250 mM monovalent salt produced a significantly higher dichroism intensity than 50 or 100 mM had (Figure 2a). The $MgSO_4$ data was more self-consistent, but still not perfectly so, particularly at 50 mM (see Figures 2b and 3b). This inconsistency with the intermediate concentrations of both monovalent and divalent cations could be due to the fact that scattering is highly sensitive to the largest particles in solution. The result was consistent with the concept that the largest particles were as shown in Figure 3a,b, but that more of these large particles form at lower salt concentrations, which resulted in more CD signal loss, possibly due at least in part to a smaller effective cross-section being available for CD measurements [32].

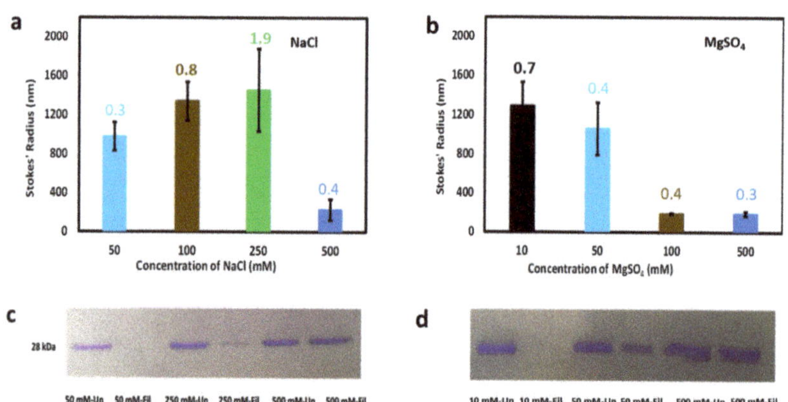

Figure 3. Particle size analysis of cl-Par-4 as a function of salt concentration. (**a**) dynamic light scattering (DLS)-derived Stokes' radii vs. NaCl concentration. (**b**) DLS-derived Stokes' radii vs. $MgSO_4$ concentration. (**c**) SDS-PAGE analysis of samples from panel (**a**), before and after filtration. (**d**) SDS-PAGE analysis of samples from panel (**b**), before and after filtration. In panels (**a**) and (**b**), error bars represent the standard deviation in Stokes' radii from multiple runs, and the number above each bar represents the average polydispersity value for these runs.

Sample filtration was next used to further analyze the relationship between CD spectroscopy and particle size. The fraction of protein removed by filtration was analyzed via SDS-PAGE. For this analysis, 0.2 mg/mL cl-Par-4 samples with low salt (50 mM NaCl or 10 mM $MgSO_4$), intermediate salt (250 mM NaCl or 50 mM $MgSO_4$), and high salt (500 mM NaCl or 500 mM $MgSO_4$) were used. Corresponding filtered samples were prepared by passing the samples through 0.45-micron filters. The results for NaCl and MgSO$_4$ samples are shown in Figure 3c,d, respectively. Each of the unfiltered samples showed similar gel band intensity. After filtration, the bands for low salt samples essentially disappeared, the bands for medium salt samplers were reduced, and the bands for high salt samples remained nearly as intense as before filtration. These results are consistent with the concept that high salt prevents large particle sizes, low salt samples consist almost entirely of large particles, and intermediate salt produces a mixture of large and small particles. In fact, Figure 3a seems to indicate that intermediate NaCl concentration produces a small number of very large particles, that are larger than the largest particles formed at low salt. One possible explanation for this observation is a sort of slow growth model: in low salt, the protein may quickly coalesce principally into 1000 nm aggregates that do not aggregate further, while the aggregates that form at medium salt are surrounded by additional smaller particles that can continue to be adsorbed by the aggregated particles, ultimately producing larger particles than seen at low salt. This continuous process would be consistent with the very high polydispersity observed for the 250 mM NaCl sample.

3.4. Effect of Filtration on DLS Measurements, CD Spectroscopy, and Secondary Structure Analysis

To further explore the relationship between particle size and CD spectroscopy, filtered samples with low, medium, and high divalent cation concentrations (10, 50, and 500 mM MgSO$_4$) were analyzed via DLS and CD spectroscopy. As expected, the hydrodynamic sizes derived from DLS measurements (Figure 4a) indicated smaller particle sizes relative to unfiltered samples (Figure 3b), except for the high salt 500 mM sample which predictably showed little change in particle size upon filtration. However, the filtered intermediate sample (50 mM MgSO$_4$) produced the largest Stokes' radius. This pattern for the filtered MgSO$_4$ samples (Figure 4a) now resembles the pattern observed for the unfiltered NaCl samples (Figure 3a), with intermediate salt producing the largest particles, possibly due to a similar slow growth mechanism as was discussed in the previous section. Bear in mind, however, that very few particles of any sort survive filtration at low salt concentration.

CD spectra of the filtered samples are shown in Figure 4b. The filtered low MgSO$_4$ sample predictably produced almost no dichroism, since the sample concentration after filtration was extremely low. As expected, the high MgSO$_4$ sample produced a CD spectrum nearly matching the pre-filtered CD spectrum of Figure 2b. The most interesting result was that for medium MgSO$_4$ (50 mM). Though the intensity was low relative to the filtered high salt sample, the shape nearly matches that of the high salt sample. This suggests that, although fewer particles were present in the filtered medium MgSO$_4$ sample (see Figure 3d), and these particles were of relatively large size (see Figure 4a), the secondary structure was very similar to that of the particles present in high salt conditions. Thus, the difference in particle size found in the filtered medium and high salt samples does not correspond to a significant change in secondary structure. Apparently, cl-Par-4 particles can self-associate without a large change in secondary structure. However, this analysis only applies to particles small enough to survive filtration. Unfiltered samples with very large particles do produce differently shaped CD spectra (Figure 2b, low and medium salt samples).

Secondary structure analysis was performed via deconvolution of the CD spectra using the SELCON3 algorithm [26]. Deconvolution of unfiltered and filtered 50 mM and 500 mM MgSO$_4$ spectra are shown in Figure 4c,d, respectively. The deconvolution program was unable to distinguish secondary structure differences between high and medium MgSO$_4$ samples, before and after filtration: in each of these four cases, deconvolution suggests approximately 80% helix, 15% disorder, and 5% beta turn. Deconvolution of the unfiltered 10 mM salt CD spectrum suggested approximately 55% helix, and corresponding increases in the percentage of beta sheet, beta turn, and disorder. However, scattering by the large particles present in this sample would be expected to interfere with CD spectroscopy and hence with spectral deconvolution, and therefore the 10 mM MgSO$_4$ secondary structure deconvolution is not presented in Figure 4c [33]. Deconvolution of the filtered low salt CD data was not possible due to lack of sufficient sample surviving filtration (see Figure 4b) and thus filtered low salt secondary structure is not presented in Figure 4d.

Both the filtered medium and high salt samples produce CD spectra with a θ_{222}: θ_{208} ratio less than one. This was consistent with the presence of mostly non-coiled coil helices in both cases. Note that before filtration, the medium salt sample produced CD spectra with a θ_{222}: θ_{208} ratio greater than one. This result suggests that the large aggregates present in the mixed-size environment of the unfiltered medium salt sample are responsible for the shifting of the θ_{222}: θ_{208} ratio, and that particles small enough to survive filtration contain mostly non-coiled coil helices.

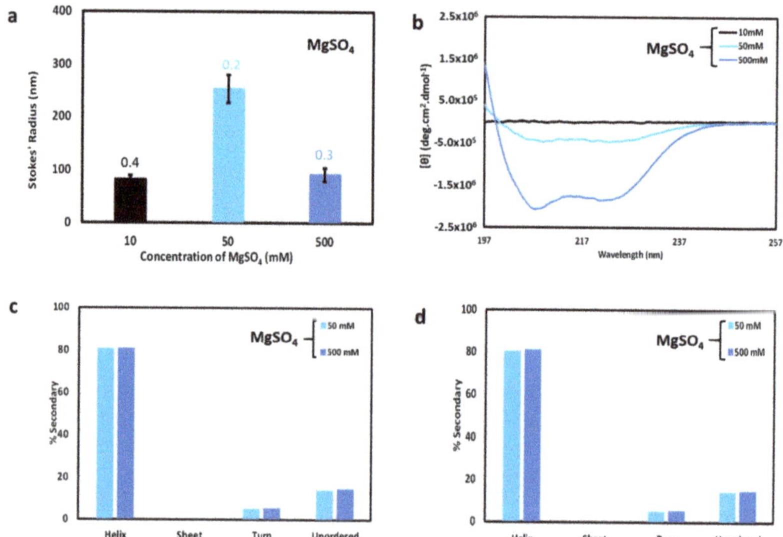

Figure 4. Effect of filtration on cl-Par-4 samples as a function of $MgSO_4$ concentration. (**a**) DLS-derived Stokes' radii after filtering. Error bars represent the standard deviation in Stokes' radii from multiple runs, and the number above each bar represents the average polydispersity value for these runs. (**b**) CD spectra after filtration. (**c**) Secondary structure analysis via deconvolution of CD spectra from Figure 2b [unfiltered samples]. (**d**) Secondary structure analysis via deconvolution of CD spectra from Figure 4b [filtered samples]. Secondary structure analysis of unfiltered and filtered 10 mM $MgSO_4$ samples was not possible due to the presence of large particles and low protein concentration, respectively.

3.5. Effect of Centrifugation and Reintroduction of Salt

To corroborate the results obtained from filtration, parallel experiments were performed by centrifuging the unfiltered samples and collecting the supernatant for CD analysis. CD spectra obtained from centrifuge supernatant samples are shown in Figure 5a. By comparison to the data of Figure 4b, it can be seen that filtration and centrifugation produced a similar effect of removing large aggregates, leaving smaller, highly helical particles in solution, although fewer particles survived filtration in the case of centrifugation. DLS analysis of the centrifuged samples used for the CD spectra in Figure 5a is shown in Figure 5b. The Stokes' radii of each of the centrifuged samples confirm that only relatively small particles survive centrifugation.

Also, to investigate the reversibility of large particle formation by cl-Par-4, 90 $MgSO_4$ was added to low salt (10 mM $MgSO_4$) samples. CD analysis showed that the resulting 100 mM $MgSO_4$ spectrum becomes nearly identical to the original 100 mM $MgSO_4$ sample which had never been at low salt (Figure 6a). Filtering of the resulting sample caused a small reduction in CD intensity, indicating the presence of a small number of large particles before filtration.

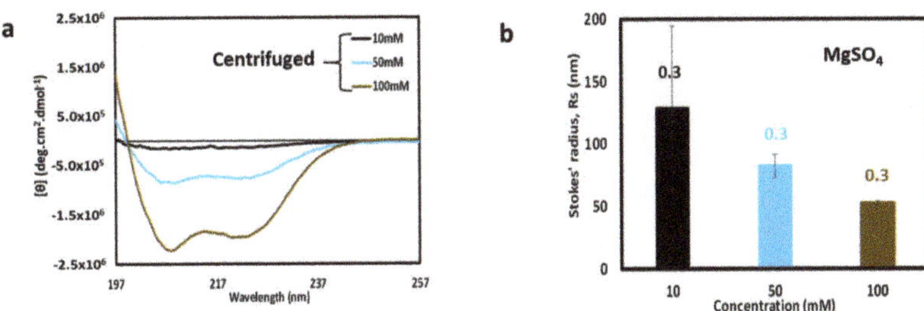

Figure 5. CD spectra and DLS analysis after centrifugation. (**a**) Overlay of CD spectra of supernatant after centrifugation of cl-Par-4 at various concentrations of $MgSO_4$. The salt concentrations are indicated by the color legend within the panel. (**b**) Corresponding DLS of samples from panel (**a**). Error bars represent the standard deviation in Stokes' radii from multiple runs, and the number above each bar represents the average polydispersity value for these runs.

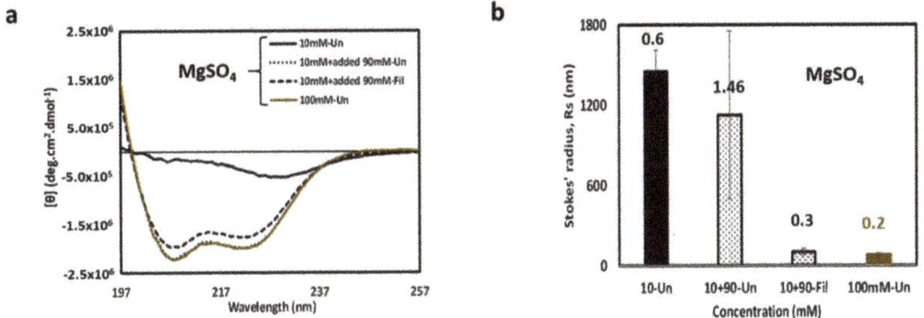

Figure 6. CD spectra and DLS analysis after reintroduction of salt. (**a**) 10 mM $MgSO_4$ sample (solid black line) shows signs of large species formation (loss of signal); This signal loss can be eliminated by increasing the $MgSO_4$ concentration to 100 mM (dotted black line); the resulting spectrum overlaps nearly perfectly with that of the original 100 mM sample (solid brown line). The dashed black line represents the slightly reduced CD spectrum of the added $MgSO_4$ sample after filtration to remove any remaining large particles. (**b**) Corresponding DLS of samples from panel (**a**). Error bars represent the standard deviation in Stokes' radii from multiple runs, and the number above each bar represents the average polydispersity value for these runs.

DLS analysis of the samples from Figure 6a is shown in Figure 6b. From left to right, the large detected Stokes' radius at 10 mM $MgSO_4$ appeared to become slightly smaller once 90 mM $MgSO_4$ was added, but the large error bar and high polydispersity indicated a wide range of particle size. After filtration, the Stokes' radius was much smaller, approaching that of the sample that had never been at low salt. The fact that filtration had a much larger effect on the DLS result than on the CD result indicates that the reconstituted 100 mM $MgSO_4$ sample contained a small number of large particles that dominate the scattering. Removal of the large particles by filtration revealed that most particles were small under these conditions.

3.6. Time-Dependent Characteristics of cl-Par-4 Samples

To further explore the possibility of dynamic equilibrium, time course CD spectroscopy experiments (Figure 7) were performed with samples at an intermediate salt concentration (50 mM $MgSO_4$). The spectra in Figure 7a,b were performed with a sample that had been filtered on day 1. The shape of the CD spectra of the filtered sample changed relatively little over a seven-day period (see Figure 7a), suggesting that the sample does not rapidly aggregate over time. However, there is some loss of intensity, suggesting either sample loss or the formation of a small number of aggregates. DLS results showed that particle sizes remained relatively small over the time course (Figure 7b), thus indicating that loss of CD intensity was most likely due primarily to sample loss.

The change in the unfiltered 50 mM $MgSO_4$ sample with time was more pronounced. The CD spectrum (Figure 7c) changed shape to resemble the filtered sample by day 3. This suggests that large particles may be lost from the unfiltered sample relatively quickly, either due to dispersion into smaller particles or via settling or adhering to the sample container. Since the intensity did not increase with time, the latter mechanism of removal by settling/adherence appeared dominant, but both mechanisms may be in play. DLS results (Figure 7d) showed a significant increase in measured Stokes' radius and dispersity until day 5, and then a reduction in size and dispersity on day 7. The maximum particle size, to which DLS is sensitive, remained relatively large throughout the time course, indicating that aggregates were present throughout. However, since the CD spectra (Figure 7c) converted over time to a shape consistent with that displayed by smaller helical particles, the combined results seems to indicate that a small number of large particles remain over time, sufficient to influence the DLS results, but not sufficient to affect the CD spectra. The overarching result is that unfiltered, filtered, and centrifuged samples converge overtime to produce similarly shaped CD spectra indicative of highly helical particles. Taken together, these data are consistent with the presence of a dynamic equilibrium between small and large particles, which readjusts the following filtration, and is affected by settling/adherence.

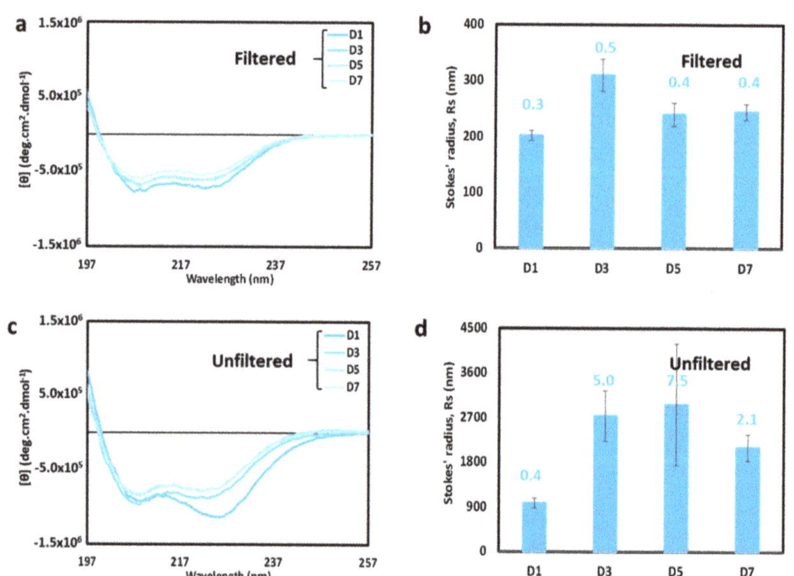

Figure 7. Time-dependence of cl-Par-4 in the presence of 50 mM $MgSO_4$. (**a**) Overlay of CD spectra obtained over a seven-day time course. Sample was filtered on day 1 prior to the first spectrum. (**b**) DLS analysis of sample from panel (**a**). (**c**) Overlay of CD spectra obtained over a seven-day time course. Sample was NOT filtered. (**d**) DLS analysis of sample from panel (**c**). In panels (**b**) and (**d**), error bars represent the standard deviation in Stokes' radii from multiple runs, and the number above each bar represents the average polydispersity value for these runs.

4. Discussion

Full-length Par-4 (FL-Par-4) has been shown to be a largely intrinsically disordered protein (IDP) in vitro: CD spectroscopy and other techniques have previously been used to show that the C-terminal coiled-coil folds in FL-Par-4, while a large majority of the remainder of the protein maintains a disordered state under neutral conditions [34]. Caspase-3-induced cleavage of FL-Par-4 to form the N-terminal 15 kilodalton PAF fragment and the C-terminal 25 kilodalton cl-Par-4 fragment is necessary for the migration of cl-Par-4 to the nucleus, where it inhibits NF-κB-mediated cell survival pathways [13,14]. The PAF fragment has been shown to function as a decoy that protects FL-Par-4 from ubiquitin-mediated degradation [35].

We have previously shown that cl-Par-4 forms small, well-folded particles with approximately 80% helical content under either high sodium or low pH conditions [23,24]. There are differences, however. High sodium results in a tetramer conformation with a Stokes' radius of approximately 100 nm, in which a significant portion of the helical content is not coiled coil. In contrast, acidic pH results in particles of a smaller size (approximately 50 nm Stokes' radius), suggestive of a dimer, which does display coiled-coil character. Here we have shown that potassium and sodium have similar effects on cl-Par-4 structure (see Figure 2a). Thus, we can state that highly helical, significantly non-coiled particles, consistent with the Stokes' radius of a tetramer, are also formed at high potassium concentration. This is a potentially meaningful finding, since potassium levels are higher inside of cells (of the order of 100 mM vs. 10 mM inside vs. outside cells), and sodium levels are higher extracellularly (of the order of 10mM and 100 mM inside vs. outside cells) [36,37]. This result suggests that cl-Par-4, which is found both intra- and extra-cellularly, may behave conformationally similarly inside and outside of the cell.

The amount of sodium or potassium necessary to induce the formation of these small helical particles appears to be higher than is typically encountered within either mammalian cellular or extra-cellular environments. In part for this reason, we investigated whether divalent cations could more efficiently induce this conformation. The results herein show that this is indeed the case, with 100 mM $MgSO_4$ inducing the formation of relatively small, highly helical particles, at concentrations approximately five times lower than the required concentrations of sodium or potassium (see Figures 2c and 3a,b). Together with previously published data identifying the conformation at high sodium concentration as a tetramer [24], this strongly suggests that tetramer is formed at the 100 mM $MgSO_4$ concentration. The nearly identical results between $MgSO_4$ and $MgCl_2$ (see Figure 2d) also show that anion identity is far less important than cation identity for affecting cl-Par-4 conformation. The required magnesium concentration is still high relative to typical intra-cellular magnesium levels [36]. However, it has been reported that apoptosis is accompanied by alteration of cellular ionic conditions, including sodium, potassium, magnesium, chloride, calcium, and zinc, and changes in pH [36,38–41]. Thus, the combined effect of a number of factors should be considered, as they may conspire to affect the conformation of cl-Par-4 and other cellular proteins under less extreme conditions than are required of any one of these factors. Furthermore, the techniques employed in these studies require high micromolar protein concentration. Since in vivo cl-Par-4 levels will not reach high micromolar concentrations, cl-Par-4 should be less prone to aggregate and may not require as high ionic strength in order to form small, well-folded particles in vivo.

In support of this view, the present analysis of CD data, DLS data including polydispersity values, sample filtration, and gel electrophoresis, has shown that a mixture of particle sizes are present under various conditions. For intermediate conditions, large particles can be removed by filtration or centrifugation, and sufficient small particles remain for secondary structure analysis. These particles, which are sufficiently small to survive filtration, apparently have a secondary structure that resembles that of the small particles found at high salt (see Figure 4b). Thus, cl-Par-4 clearly can form highly helical structures under both low and medium salt conditions. This is at least true for particles that are not highly aggregated. Secondary structure analysis of the larger aggregates is complicated by the effects of scattering and will be discussed in more detail elsewhere.

As mentioned, approximately 80% helicity is observed in the small cl-Par-4 particles induced by sodium, potassium, or magnesium. The coiled-coil domain comprises only 38% of the cl-Par-4 sequence (see Figure 1a), thus additional helical regions are clearly present. Sequence analysis shows that the SAC domain, which comprises 28% of cl-Par-4, has the next highest order propensity (see Figure 1b). The linker domain has the highest disorder propensity. This strongly suggests that both the CC domain and at least a significant fraction of the SAC domain are helical in cl-Par-4 under these conditions. Previous biophysical analysis of FL-Par-4 provided no such evidence of folding of the SAC domain in the full-length protein [34] and suggested that the helical content in FL-Par-4 is largely coiled-coil in nature. Thus, cleavage of Par-4 by caspase-3 may be important both for altering the SAC domain conformation, and for influencing the self-association characteristics, including helical coiling, of the cl-Par-4 fragment in comparison to FL-Par-4. These two effects may in fact be correlated. It remains to be seen how conformational and self-association changes upon caspase-induced cleavage may be related to controlling the role of Par-4 in apoptotic processes.

Author Contributions: Conceptualization, S.M.P.; methodology, K.K.R. and S.M.P.; formal analysis, K.K.R. and S.M.P.; investigation, K.K.R., K.P. and S.M.P.; resources, S.M.P.; writing—original draft preparation, K.K.R.; writing—review and editing, S.M.P. and K.K.R.; visualization, K.K.R. and S.M.P.; supervision, S.M.P.; project administration, S.M.P.; funding acquisition, S.M.P. All authors have read and agreed to the published version of the manuscript.

Funding: This research was supported by internal funds from Old Dominion University.

Institutional Review Board Statement: Not applicable.

Informed Consent Statement: Not applicable.

Acknowledgments: We wish to acknowledge Vivek Rangnekar for useful discussions regarding Par-4 function.

Conflicts of Interest: The authors declare no conflict of interest.

References

1. El-Guendy, N.; Zhao, Y.; Gurumurthy, S.; Burikhanov, R.; Rangnekar, V.M. Identification of a Unique Core Domain of Par-4 Sufficient for Selective Apoptosis Induction in Cancer Cells. *Mol. Cell Biol.* **2003**, *23*, 5516–5525. [CrossRef]
2. Rangnekar, V.M. Apoptosis by par-4 protein. In *Advances in Cell Aging and Gerontology*; Elsevier: Amsterdam, The Netherlands, 2001; pp. 215–236.
3. Boghaert, E.R.; Sells, S.F.; Walid, A.J.; Malone, P.; Williams, N.M.; Weinstein, M.H.; Strange, R.; Rangnekar, V.M. Immunohistochemical analysis of the proapoptotic protein Par-4 in normal rat tissues. *Cell Growth Differ.* **1997**, *8*, 881–890.
4. Gurumurthy, S.; Goswami, A.; Vasudevan, K.M.; Rangnekar, V.M. Phosphorylation of Par-4 by protein kinase A is critical for apoptosis. *Mol. Cell Biol.* **2005**, *25*, 1146–1161.
5. Rasool, R.U.; Nayak, D.; Chakraborty, S.; Katoch, A.; Faheem, M.M.; Amin, H.; Goswami, A. A journey beyond apoptosis: New enigma of controlling metastasis by pro-apoptotic Par-4. *Clin. Exp. Metastasis* **2016**, *33*, 757–764. [CrossRef] [PubMed]
6. Johnstone, R.W.; Tommerup, N.; Hansen, C.; Vissing, H.; Shi, Y. Mapping of the human PAWR (par-4) gene to chromosome 12q21. *Genomics* **1998**, *53*, 241–243. [CrossRef] [PubMed]
7. Treude, F.; Kappes, F.; Fahrenkamp, D.; Muller-Newen, G.; Dajas-Bailador, F.; Kramer, O.H.; Luscher, B.; Hartkamp, J. Caspase-8-mediated PAR-4 cleavage is required for TNFalpha-induced apoptosis. *Oncotarget* **2014**, *5*, 2988–2998. [CrossRef] [PubMed]
8. Thayyullathil, F.; Pallichankandy, S.; Rahman, A.; Kizhakkayil, J.; Chathoth, S.; Patel, M.; Galadari, S. Caspase-3 mediated release of SAC domain containing fragment from Par-4 is necessary for the sphingosine-induced apoptosis in Jurkat cells. *J. Mol. Signal.* **2013**, *8*, 2. [CrossRef] [PubMed]
9. El-Guendy, N.; Rangnekar, V.M. Apoptosis by Par-4 in cancer and neurodegenerative diseases. *Exp. Cell Res.* **2003**, *283*, 51–66. [CrossRef]
10. Burikhanov, R.; Zhao, Y.; Goswami, A.; Qiu, S.; Schwarze, S.R.; Rangnekar, V.M. The tumor suppressor Par-4 activates an extrinsic pathway for apoptosis. *Cell* **2009**, *138*, 377–388. [CrossRef] [PubMed]
11. Hebbar, N.; Wang, C.; Rangnekar, V.M. Mechanisms of apoptosis by the tumor suppressor Par-4. *J. Cell Physiol.* **2012**, *227*, 3715–3721. [CrossRef]
12. de Thonel, A.; Hazoumé, A.; Kochin, V.; Isoniemi, K.; Jego, G.; Fourmaux, E.; Hammann, A.; Mjahed, H.; Filhol, O.; Micheau, O.; et al. Regulation of the proapoptotic functions of prostate apoptosis response-4 (Par-4) by casein kinase 2 in prostate cancer cells. *Cell Death Dis.* **2014**, *5*, e1016. [CrossRef]

13. Chaudhry, P.; Singh, M.; Parent, S.; Asselin, E. Prostate apoptosis response 4 (Par-4), a novel substrate of caspase-3 during apoptosis activation. *Mol. Cell Biol.* **2012**, *32*, 826–839. [CrossRef] [PubMed]
14. Diaz-Meco, M.T.; Lallena, M.J.; Monjas, A.; Frutos, S.; Moscat, J. Inactivation of the inhibitory kappaB protein kinase/nuclear factor kappaB pathway by Par-4 expression potentiates tumor necrosis factor alpha-induced apoptosis. *J. Biol. Chem.* **1999**, *274*, 19606–19612. [CrossRef] [PubMed]
15. Shrestha-Bhattarai, T.; Rangnekar, V.M. Cancer-selective apoptotic effects of extracellular and intracellular Par-4. *Oncogene* **2010**, *29*, 3873–3880. [CrossRef] [PubMed]
16. Uversky, V.N. Intrinsically Disordered Proteins and Their "Mysterious" (Meta)Physics. *Front. Phys.* **2019**, *7*, 10. [CrossRef]
17. Wright, P.E.; Dyson, H.J. Intrinsically disordered proteins in cellular signalling and regulation. *Nat. Rev. Mol. Cell Biol.* **2015**, *16*, 18–29. [CrossRef]
18. Dyson, H.J.; Wright, P.E. Intrinsically unstructured proteins and their functions. *Nat. Rev. Mol. Cell Biol.* **2005**, *6*, 197–208. [CrossRef]
19. Uversky, V.N.; Gillespie, J.R.; Fink, A.L. Why are "natively unfolded" proteins unstructured under physiologic conditions? *Proteins* **2000**, *41*, 415–427. [CrossRef]
20. Uversky, V.N. What does it mean to be natively unfolded? *Eur. J. Biochem.* **2002**, *269*, 2–12. [CrossRef]
21. Uversky, V.N. Intrinsically disordered proteins and their environment: Effects of strong denaturants, temperature, pH, counter ions, membranes, binding partners, osmolytes, and macromolecular crowding. *Protein J.* **2009**, *28*, 305–325. [CrossRef]
22. Wicky, B.I.M.; Shammas, S.L.; Clarke, J. Affinity of IDPs to their targets is modulated by ion-specific changes in kinetics and residual structure. *Proc. Natl. Acad. Sci. USA.* **2017**, *114*, 9882–9887. [CrossRef] [PubMed]
23. Clark, A.M.; Ponniah, K.; Warden, M.S.; Raitt, E.M.; Yawn, A.C.; Pascal, S.M. pH-Induced Folding of the Caspase-Cleaved Par-4 Tumor Suppressor: Evidence of Structure Outside of the Coiled Coil Domain. *Biomolecules* **2018**, *8*, 4060–4073. [CrossRef] [PubMed]
24. Clark, A.M.; Ponniah, K.; Warden, M.S.; Raitt, E.M.; Smith, B.G.; Pascal, S.M. Tetramer formation by the caspase-activated fragment of the Par-4 tumor suppressor. *FEBS J.* **2019**, *286*, 4060–4073. [CrossRef]
25. Alexandrov, A.; Dutta, K.; Pascal, S.M. MBP fusion protein with a viral protease cleavage site: One-step cleavage/purification of insoluble proteins. *Biotechniques* **2001**, *30*, 1194–1198. [CrossRef]
26. Whitmore, L.; Wallace, B.A. DICHROWEB, an online server for protein secondary structure analyses from circular dichroism spectroscopic data. *Nucleic Acids Res.* **2004**, *32*, W668–W673. [CrossRef] [PubMed]
27. Ward, J.J.; McGuffin, L.J.; Bryson, K.; Buxton, B.F.; Jones, D.T. The DISOPRED server for the prediction of protein disorder. *Bioinformatics* **2004**, *20*, 2138–2139. [CrossRef]
28. Jones, D.T.; Cozzetto, D. DISOPRED3: Precise disordered region predictions with annotated protein-binding activity. *Bioinformatics* **2014**, *31*, 857–863. [CrossRef]
29. Wuo, M.G.; Mahon, A.B.; Arora, P.S. An Effective Strategy for Stabilizing Minimal Coiled Coil Mimetics. *J. Am. Chem. Soc.* **2015**, *137*, 11618–11621. [CrossRef]
30. Kallenbach, N.R.; Lyu, P.; Zhou, H. CD Spectroscopy and the Helix-Coil Transition in Peptides and Polypeptides. In *Circular Dichroism and the Conformational Analysis of Biomolecules*; Fasman, G.D., Ed.; Springer: Boston, MA, USA, 1996; pp. 201–259.
31. Cooper, T.M.; Woody, R.W. The effect of conformation on the CD of interacting helices: A theoretical study of tropomyosin. *Biopolymers* **1990**, *30*, 657–676. [CrossRef]
32. Wallace, B.A.; Mao, D. Circular dichroism analyses of membrane proteins: An examination of differential light scattering and absorption flattening effects in large membrane vesicles and membrane sheets. *Anal. Biochem.* **1984**, *142*, 317–328. [CrossRef]
33. Kelly, S.M.; Jess, T.J.; Price, N.C. How to study proteins by circular dichroism. *Biochim. Biophys. Acta Proteins Proteom.* **2005**, *1751*, 119–139. [CrossRef] [PubMed]
34. Libich, D.S.; Schwalbe, M.; Kate, S.; Venugopal, H.; Claridge, J.K.; Edwards, P.J.B.; Dutta, K.; Pascal, S.M. Intrinsic disorder and coiled-coil formation in prostate apoptosis response factor 4. *FEBS J.* **2009**, *276*, 3710–3728. [CrossRef] [PubMed]
35. Hebbar, N.; Burikhanov, R.; Shukla, N.; Qiu, S.; Zhao, Y.; Elenitoba-Johnson, K.S.J.; Rangnekar, V.M. A Naturally Generated Decoy of the Prostate Apoptosis Response-4 Protein Overcomes Therapy Resistance in Tumors. *Cancer Res.* **2017**, *77*, 4039–4050. [CrossRef]
36. Ishaque, A.; Al-Rubeai, M. Role of Ca2+, Mg2+ and K+ ions in determining apoptosis and extent of suppression afforded by bcl-2 during hybridoma cell culture. *Apoptosis* **1999**, *4*, 335–355. [CrossRef] [PubMed]
37. Melkikh, A.V.; Sutormina, M.I. Model of active transport of ions in cardiac cell. *J. Theor. Biol.* **2008**, *252*, 247–254. [CrossRef]
38. Pilchova, I.; Klacanova, K.; Tatarkova, Z.; Kaplan, P.; Racay, P. The Involvement of Mg(2+) in Regulation of Cellular and Mitochondrial Functions. *Oxid. Med. Cell Longev.* **2017**, *2017*, 6797460. [CrossRef]
39. Franklin, R.B.; Costello, L.C. The important role of the apoptotic effects of zinc in the development of cancers. *J. Cell Biochem.* **2009**, *106*, 750–757. [CrossRef]
40. Bortner, C.D.; Cidlowski, J.A. Ion channels and apoptosis in cancer. *Philos. Trans. R Soc. B* **2014**, *369*, 20130104. [CrossRef]
41. Lang, F.; Föller, M.; Lang, K.S.; Lang, P.A.; Ritter, M.; Gulbins, E.; Vereninov, A.; Huber, S.M. Ion Channels in Cell Proliferation and Apoptotic Cell Death. *J. Membr. Biol.* **2005**, *205*, 147–157. [CrossRef]

MDPI
St. Alban-Anlage 66
4052 Basel
Switzerland
Tel. +41 61 683 77 34
Fax +41 61 302 89 18
www.mdpi.com

Biomolecules Editorial Office
E-mail: biomolecules@mdpi.com
www.mdpi.com/journal/biomolecules